New Connectivities in China

Pui-lam Law
Editor

New Connectivities in China

Virtual, Actual and Local Interactions

Springer

Editor
Pui-lam Law
Department of Applied Social Sciences
The Hong Kong Polytechnic University
Hong Kong, China

ISBN 978-94-007-3909-3 ISBN 978-94-007-3910-9 (eBook)
DOI 10.1007/978-94-007-3910-9
Springer Dordrecht Heidelberg London New York

Library of Congress Control Number: 2012935696

Printed on acid-free paper

Springer is part of Springer Science+Business Media (www.springer.com)

Acknowledgments

I gratefully acknowledge permission of the publishers to reprint the following articles. Yang, B. (2009). NPOs in China: Some issues concerning Internet communication. *Knowledge, Technology and Policy*, 21(1), 37–42; Fortunati, L., Manganelli, A. M., Law, P., & Yang, S. (2008). Beijing calling… Mobile communication in contemporary China. *Knowledge, Technology and Policy*, 21(1), 19–27; Fortunati, L., Manganelli, A. M., Law, P., & Yang, S. (2008). The mobile face of contemporary China. *Telektronikk*, 104(2), 57–67; Yang, K. (2008). A preliminary study on the use of mobile phones amongst migrant workers in Beijing. *Knowledge, Technology and Policy*, 21(2), 65–72; Lin, A., & Tong, A. (2008). Mobile cultures of migrant workers in southern China: Informal literacies in the negotiation of (new) social relations of the new working women. *Knowledge, Technology and Policy*, 21(2), 73–81; Peng, Y. (2008). Internet use of migrant workers in the Pearl River Delta. *Knowledge, Technology and Policy*, 21(2), 47–54; Ngan, R., & Ma, S. (2008). The relationship of mobile telephony to job mobility in China's Pearl River Delta. *Knowledge, Technology and Policy*, 21(2), 55–63; Chu, W. C. R. (2008). The dynamics of cyber China: The characteristics of Chinese ICT use. *Knowledge, Technology and Policy*, 21(1), 29–35 (retitled in Chapter 16 as "ICT Use with Chinese Characteristics" and co-authored with Y. Peng); Zhao, J. (2008). ICT4D: Internet adoption and usage among rural users in China. *Knowledge, Technology and Policy*, 21(1), 9–18.

I would like to thank the authors for their relentless commitment and their contributions to this collection. Finally, I would also like to express our gratitude to the reviewers for their thoughtful and meticulous comments.

Contents

Notes on Contributors

Michael Arnold is senior lecturer in the History and Philosophy of Science Programme in the School of Historical and Philosophical Studies at the University of Melbourne, Australia. His ongoing teaching and research activities lie at the intersection of contemporary technologies and our society and culture. Michael's currently funded research projects include (1) a comparative study of social networking in six locations across the Asia-Pacific, (2) a longitudinal examination of high-speed broadband in the domestic context, (3) a study to develop methods to assess social returns on investments in technologies, (4) a study of ethical and governance issues associated with the electronic health record, and (5) a study of online memorials. Michael is a founding committee member and current Secretary of the IEEE Society on Social Implications of Technology (Australian Chapter) and is an Australian committee member of the Community Informatics Research Network.

Chung-tai Cheng gained his Ph.D. in Sociology from the Peking University in China. He is currently research associate of the Centre for Third Sector Studies in the Department of Applied Social Sciences, The Hong Kong Polytechnic University. His research interest is on the topic of the normative implications of ICT uses in China. His journal articles and book chapters are mainly on social and cultural impacts of ICTs.

Wai-chi Rodney Chu obtained his Ph.D. in Sociology from The University of Queensland, Australia. He is assistant professor in the Department of Applied Social Sciences, The Hong Kong Polytechnic University. Currently, he is focusing on research concerning the sociocultural impact of ICTs on contemporary Chinese, with particular reference to the use of mobile telephony and the Internet by migrant workers and university students.

Leopoldina Fortunati is professor of sociology of communication and sociology of cultural processes in the Faculty of Educational Sciences at the University of Udine, Italy. She has conducted several researches in the field of gender studies, cultural processes, and communication technologies. She is the author of many books and has published extensively in important journals. She represents Italy in the COST Domain Committee "Individuals, Societies, Cultures and Health" and in the COST 298 "Participation in the Broadband Society." She is the co-chair of the international association "The Society for the Social Study of Mobile Communication" (SSSMC) which aims to facilitate the international advancement of cross-disciplinary mobile communication studies. Her work has been published in 11 languages.

David Kurt Herold taught and researched in China for over 9 years, before joining the Hong Kong Polytechnic University as lecturer for sociology. His research is focused on the use of ICTs by humans. In particular, he studies the Chinese Internet, encounters between Chinese and non-Chinese online, the impact of the internet on offline society, and online education. His recent publications include Cultural Politics and Political Culture of Web 2.0 in Asia (2009), Knowledge, Technology, and Policy; Mediating Media Studies—Stimulating Critical Awareness in a virtual environment (2010), Computers & Education; Imperfect Use? ICT Provisions and Human Decisions (2010), The Information Society; and Online Society in China (2011), Routledge.

Larissa Hjorth is an artist, digital ethnographer, and senior lecturer in the Games Programs, School of Media and Communication, RMIT University, Australia. Since 2000, Hjorth has been researching and publishing on gendered mobile media, gaming, and virtual communities in the Asia-Pacific—these studies are outlined in her book, *Mobile Media in the Asia-Pacific* (London, Routledge). Hjorth has published widely on the topic in national and International journals such as *Games and Culture Journal*, *Convergence Journal*, *Journal of Intercultural Studies*, *Continuum*, *ACCESS*, *Fibreculture* and *Southern Review*, and in 2009 coedited two Routledge anthologies, *Gaming Cultures and Place in the Asia-Pacific Region* (with Dean Chan) and *Mobile Technologies: From Telecommunication to Media* (with Gerard Goggin). In 2010, Hjorth released *Games & Gaming* textbook (London: Berg).

Pui-lam Law received his Ph.D. in sociology from The University of New South Wales, Australia. He is currently assistant professor in the Department of Applied Social Sciences of The Hong Kong Polytechnic University and visiting research fellow at the Centre for Creative Industries Studies of Peking University. His research interest is on modernity and social development in China. Currently, he is working on the ICTs and the issue of migrant workers identity in southern China.

Jing Li obtained her Masters degree in sociology at Peking University, Beijing, China, in 2007. Currently, she is the head of human resources office in Economic Consulting Center of the State Information Center. She has studied the mobile telephony and migrant workers and the ICTs and social stratification since 2004 and published journal articles and book chapters on the social consequence of the ICTs.

Angel Mei-yi Lin received her Ph.D. from the Ontario Institute for Studies in Education, University of Toronto, Canada, in 1996. She is currently an associate professor in the Faculty of Education, The University of Hong Kong. She has been an active and productive researcher in the areas of critical discourse analysis, urban ethnography, critical pedagogy, feminist media studies, and youth literacies and cultural studies. With a background in ethnomethodology, conversation analysis, and social theory, her theoretical orientations are phenomenological, sociocultural, and critical. In recent years, she has researched mobile SMS cultures of young people in Hong Kong and China, and has published in international research journals including the *Journal of Gender Studies, Discourse and Society,* and *Continuum: Journal of Media and Cultural Studies*.

Stephen K. Ma is currently lecturer of the Department of Applied Social Studies, City University of Hong Kong. He got a M.A. in Sociology from the University of Kent at Canterbury, UK. He is an expert in social research, program evaluation, and statistical analysis. He has been involving in a large number of empirical studies on social research over the past 10 years.

Anna Maria Manganelli is Professor of Social Psychology at University of Padova, Department of General Psychology. She has carried out studies and researches in the following fields: (1) methodology of psychosocial research: construction of measurement instruments for social and psychosocial research techniques and methodologies to evaluate the psychometric properties of such instruments; (2) social psychology: causal attribution, "modern" forms of prejudice and their relationships with political orientation and attitudes toward immigration; (3) economic psychology: test of the Theory of Reasoned Action and successive models in the formation of saving intention.

Raymond Ngan is an associate professor at the Department of Applied Social Studies, City University of Hong Kong. He had been the former Chairman of the Committee on Social Security, Hong Kong Council of Social Service, responsible for the First (1996) and Second (2000) Asia Regional Conference on Social Security, and the Poverty Summit (1998). His research interests are in social protection for migrant workers in China, reforms in old age pension, and long-term care. His work has appeared in international journals, notably Social Development Issues, International Journal of Social Welfare, International Journal of Employment Studies, International Journal of Public Opinion Research, Social Security (Beijing), and Guangdong Social Insurance Bulletin.

Yinni Peng obtained her B.A. (2002) and M.A. (2005) in Sociology from Peking University, and received her Ph.D. in Sociology from the Chinese University of Hong Kong (2009). She was awarded J. William Fulbright Scholarship in 2008–2009. She was a post-doctoral fellow in the Department of Sociology of Hong Kong Baptist University in 2009–2010. Currently, she is assistant professor in Hong Kong Shue Yan University. She is interested in migration, labor politics, and the Information and Communication Technologies. Her journal articles and book chapters mainly focus on the social consequences of ICTs usage in contemporary mainland China.

Matteo Tarantino is an Adjunct Research Fellow at the Catholic University of Milan. His research interests focus on the nexus between culture and computer technology, especially in non-western contexts; and on the relationship between ICTs and urban environments.

Avin H.M. Tong was a research associate in the Faculty of Education, the Chinese University of Hong Kong. She received her master's degree in Communication from the Chinese University of Hong Kong. She has been researching on topics in youth cultures, media audience studies, and feminist cultural studies. She is currently a freelance writer and researcher.

Boxu Yang received his Ph.D. in Sociology from the University of Toronto, Canada, and is currently professor at the School of Journalism and Communication, Peking University. He is one of the pioneers who systemically examined the implications of the processes of globalization and glocalization in China. His current research includes the social effect of BBS forum, SNS communication, and the social interaction on the Chinese Internet. He has published in national and international journals on the topics such as international and global communication, social networks and local communication, and spaces and communications in China.

Ke Yang is a Ph.D. candidate of the Department of Sociology of Peking University and also an assistant professor at the Institute of Sociology of Chinese Academy of Social Sciences. She has been focusing her research work at the urbanization of modern China, and her main concern is the action of subaltern groups in response to the huge change of society.

Shanhua Yang is professor in the Department of Sociology at Peking University. His teaching and research interests focus on sociology of family, rural and urban sociology, and sociological theory. He is currently conducting research on family in contemporary China, the notion of trust, and the ICTs and social stratification in China. He has published extensively in national journals. His books are mainly on Chinese family structure and Chinese rural society. Recently, he is coediting a book on mobile phone and China.

Jinqiu Zhao is an associate professor at the Institute of Communication Studies at Communication University of China in Beijing. She received her Ph.D. from the Wee Kim Wee School of Communication and Information at Nanyang Technological University in Singapore. She has been a university teacher and researcher since 1993 and has published articles on various aspects of Asian communication studies in journals and books around the world. Her research interests include issues in Chinese communication, comparative media systems, social impact of new communication technologies, and media effects.

Introduction

Pui-lam Law

The speed of penetration of the information and communication technologies (ICTs) in China has been astounding in recent years. China had only 300 mobile phone users in China in 1987; by 1997, this figure had increased to 13 million Chinese (1.05% of the population), while by early 2010, 786.5 million Chinese had a mobile phone. The penetration of the Internet has likewise been dramatic. The number of users had increased from 0.62 million in 1997 to 162 million in mid-2007. A recent report delivered by the China Internet Network Center in July 2010 indicated that the number of Internet users had reached 420 million.

The fast and ubiquitous diffusion of the ICTs has brought forth new forms of connection among the Chinese. Nowadays, parents living in the hinterland are able to use an MSN Webcam in an Internet cafe to chat with their son who might be working in Shenzhen, the first special economic zone in China. Migrant workers can connect with their kinsmen or fellow villagers through the Internet or the mobile phone even when they are floating around the coastal cities. Chinese are now enjoying a higher degree of autonomy as they can get information from, and express their views on, local or international websites. These new connectivities have changed considerably the social lives of Chinese. Virtual networks have been developed either through the social networking system or the instant messaging system. These kinds of network do not operate only in the virtual space; some have led to the formation of networks in the actual world. A recent incident in which the Cantonese dialect was defended in Guangzhou city is an example to illustrate how the virtual connection has led to actual interaction: the protest in the virtual world facilitated the protest in the actual world. This is indeed common around the globe. Incidents such as this have strengthened the belief that the effects on the virtual and actual worlds brought about by the new connectivities in Chinese society would be similar to other global societies. It is no wonder that some believe the new connectivities will enhance the development of the public sphere in the actual world in China. This idea represents the thought that the globalized culture will uproot the local culture via the penetration of the ICTs. There are, however, cases which present different representations. Interactions in the actual world may well be influenced by the virtual connection, but some may also be shaped by the local culture and in turn influence the virtual interactions. The fervent nature of Chinese nationalism has had a strong influence on the virtual interactions. The local interactions may shape the interactions in the actual and then virtual worlds.

Without doubt, the new connectivities brought about by the adoption of the ICTs in China have led to complications in the relationship between virtual, actual, and local interactions. This collection attempts to elaborate various aspects of the implications of the new connectivities on these three types of interactions in China. The collection is divided into five parts. The first part explores the possibility of the development of civil society in China; the second reports the research results of the mobile phone study in Beijing, the capital city of China; the third examines the implications of the new connectivities on the migrant workers in the southern part of China; the fourth investigates how the new connectivities pose a challenge to the traditional social order and how they develop new social networks both in the virtual and the actual worlds; the final part delineates the relationship between the new connectivities and the Chinese social context.

The Internet Communication and the Issue of Civil Society

Part I mainly focuses on the relationship between the penetration of the Internet and the possibility of the development of civil society in China. The speed of the penetration of the information and communication technologies (ICTs) has been astounding in recent years. China had more Internet users than any other country in 2009, and the Internet has become one of the most important parts in the lives of Chinese people. The Internet enables them to access a wider range of news and express themselves more freely. Some scholars, taking an optimistic attitude, maintain that the Internet can facilitate the development of civil society in China by revitalizing a Habermasian sense of the public sphere on the Internet. They believe that Internet communication enables the grassroots to participate in the political, social, and cultural spheres because the platforms provided are open to all and inherently interactive. In addition, the users are able to maintain their anonymity and this is particularly important in Chinese society.

But the chapters in this section tell different stories. In "NPOs in China: Some Issues Concerning Internet Communication," Boxu Yang maintains that in theory Internet communication is well-suited to the purposes of non-profit organizations (NPOs), which are similar to the non-government organizations in the West. He points out that many Chinese who were once passive observers have been transformed into active participants as a result of Internet communication. This is significant because a civil society in China can only be achieved through the efforts of agents, not observers. Internet users have been actively learning the goals of civil society and participating in civil and political activities on the Internet. But he further elaborates that, in practice, Chinese NPOs do not take full advantage of the Internet because of their bureaucratic and elitist approach and hence Chinese NPOs rarely use the medium for advocacy. In general, Yang's chapter suggests that the interaction of the virtual cannot transform the interaction of the actual because of the influence of traditional cultural forces.

In Chap. 2, “Dot the “I’s” and Cross the “T’s”: A Sociological Interpretation of Chinese Cyberspace and the Qianyuan Grace Wang Incident,” Chung-tai Cheng also points out that the Internet may be unable to facilitate a rational discussion in both the virtual and the actual worlds and that the presentation and interpretation of the Internet are co-constructed and influenced by indigenous Chinese culture. By drawing on a case study of the “Qianyuan Grace Wang Incident” in China during the spring of 2008, Cheng explores how the power of the Internet reframes and reconstructs people’s understanding of, interpretation of, and participation in events. But he further argues that the operation of the discussion of the Internet in the Chinese societies may distort the social meaning of the online public sphere even though the Internet is regarded as an emerging platform for freedom of speech in China. He concludes that the Internet in China in this present form can hardly be a platform for rational debate; on the other hand, the presentation and the interpretation of the Internet are inherently influenced by indigenous Chinese culture. Cheng’s chapter reflects the idea that the actual has seriously affected the development of the virtual.

David Herold in Chap. 3 also discusses the negotiation among the virtual, actual, and local. He maintains that the local interaction presented through the virtual has far-reaching implications for China’s future relationship with the world. Chinese Nationalism is a good example. Chinese Nationalism online has been the topic of much recent debate both in academic circles and in the media. During 2007 and 2008, a number of topics incited Chinese netizens in China to make proclamations of nationalistic pride and to make accusations of treason against Chinese who were not “Chinese” enough, as well as running campaigns against non-Chinese who were seen as “attacking China.” In “Rage and Reflection: Chinese Nationalism Online Between Emotional Venting and Measured Opinion,” Herold argues that while many of the postings online displayed a serious lack of knowledge and understanding of China’s history and of its relationship with the world, not all of the online debates can be labeled as the emotional venting by China’s *fenqing* (愤青angry youth). He further maintains that there are a number of more reflective voices whose well-argued nationalism is far more problematic for China’s future relationship with other countries despite the fact that much of the Chinese nationalism online can be interpreted as a sign of frustration among young Chinese.

In view of the recent developments presented in these three chapters, it seems that the idea that the Internet in China will facilitate the development of civil society cannot be easily substantiated. In this aspect, in China, it seems that the virtual cannot easily transform the actual when facing the powerful countervailing force of the local.

Studies on Mobile Phone Uses in Beijing

The number of mobile users in China had reached 786.5 million by early 2010, an almost 55-times increase in a decade. The penetration rate is highest in the economically well-developed coastal cities and is close to 100% in Beijing and

Shanghai. Part II discusses the study of the social consequences of the mobile phone penetration in Beijing, and focuses on how, after a decade of mobile phone use, the inhabitants of Beijing evaluate the changes in the social and communicative sphere as a result of the introduction of the mobile phone. While a considerable amount of qualitative research devoted to ICTs in China is already available, quantitative studies are lacking. In order to fill this gap, a survey was conducted in Beijing in 2006. Chapters 4 and 5 by Leopoldina Fortunati, Anna Maria Manganelli, Pui-lam Law, and Shanhua Yang present some results of the survey specifically focused on mobile communication. Based on face-to-face questionnaires administered to a convenience sample of 487 respondents, the first chapter "Beijing Calling… Mobile Communication in Contemporary China" addresses the following research questions: After a decade of mobile phone use, how do Chinese people perceive the importance of this device? To what extent do the adoption and use of mobile phones increase or decrease social connectivity in contemporary China? To what extent does the use of mobile phones in everyday life enhance or reduce the communications—do they make users feel closer to or more distant from other people? And what are the variables that predict users' attitudes toward mobile phones in China? The second chapter, "The "Mobile" Face of Contemporary China," attempts to answer the following research question: How is the relational sphere in China being reshaped by the massive use of the mobile phone? And are there striking differences between the attitudes, behaviors, and practices associated with mobile phone use in China and in the West? The results of this study unravel the effect of the mobile phone uses on the traditional social interaction in the actual world. Thus, the data presented here will provide direction for further inquiries into various aspects of the new connectivities in China.

Chapter 6 presents the data of in-depth interviews with migrant workers in Beijing, offering a supplement to the quantitative survey. In "A Preliminary Study on the Mobile Phone Use of Migrant Workers in Beijing," Ke Yang discusses the implications of mobile phone use among migrant workers in Beijing. Through telling the stories of the workers' daily lives, Yang explicates their mobile phone usage against their alienating background in Beijing. She explores the responses of the migrants toward their situation in relation to the social usage of the mobile phone and specifically introduces and discusses four concepts for understanding patterns of migrant mobile phone use: feigned presence, concern in absence, *jianghu* relations, and romantic relations. The discussion of these four concepts further illuminates the effects of the new connectivities on virtual, actual, and local interactions.

The ICTs and Migrant Workers in Southern China

While Ke Yang in her chapter presents the effects of the mobile phone uses on the migrant workers in Beijing, Part III explores the relationships between the new connectivities and migrant workers in the southern part of China. Since the early 1980s, when China implemented economic reforms and opened its doors to the outside world, the rate and scale of the internal migration from rural to urban areas and from the western and central regions of the country to the eastern coastal region has been incessantly increasing. The floating population, predominantly migrant workers, has increased sharply from 30 million in the early 1980s to 225 million in the early 2008. This scale of internal migration is unprecedented. The speed of the penetration of the devices for the new connectivities, such as the mobile phone and the Internet, among migrant workers in recent years is also astonishing. The penetration rate of these devices is highest in the coastal cities. In late 2007, Shanghai had an almost 100% mobile penetration rate and Guangdong, 84.3%. These two cities are densely populated with migrant workers. Undoubtedly, this penetration has had a significant impact on migrant workers' social lives, in actual and virtual worlds, both directly and indirectly. In "Mobile Cultures of Migrant Workers in Southern China: Informal Literacies in the Negotiation of (New) Social Relations of the New Working Women," Lin and Tong analyze the data collected through in-depth interviews with migrant workers in Southern China about their mobile cultures. In particular, they focus on understanding the role that mobile cultures play in female workers' negotiation of their social and romantic relations and leisure space, and how these negotiations are directly or indirectly facilitated by the development of informal literacies through their frequent SMS communicative practices. This aids understanding of the lifestyle aspirations and life trajectories of the new young working women in China, who are experiencing the most rapid socioeconomic changes in society and negotiating their ways of life amidst much tension between old and new values governing lifestyle aspirations, and familial and gender relations.

Chapter 8 is "Internet Use of Migrant Workers in the Pearl River Delta." First, Yinni Peng illustrates that educational attainment is a significant factor in determining the type of Internet use that appeals to migrant workers. Second, she points out that both work-oriented and entertainment-oriented uses of the Internet have a positive influence on migrant workers' jobs and social lives in the Pearl River Delta. Access to work-related information improves their employment opportunities, while the various means of entertainment provide relaxation. The online chat and the virtual community provide migrant workers with emotional support and a strong sense of their local identity. The Internet café and online games allow migrant workers to relieve the pressures of work and passively challenge the power hierarchy in the modern factory regime. And she concludes that the uses of the new connectivities have become an integral part of the actual and virtual lives of migrant workers in the Pearl River Delta.

As the new connectivities have become central to the city lives of the migrant workers, so their impact on the local interactions of the workers has grown. Ngan and Ma in Chap. 9 hypothesize the possible relationship between mobile telephony and job mobility among migrants. They conducted a study of 655 migrant workers in 2006 in the PRD and found a relationship between the rate of job change among migrant workers and the use of the mobile phone. This is simply because of the fact that more information on jobs with better pay and working conditions can be sent swiftly through the mobile phone by friends, former coworkers, and kinsmen. Presumably, this relationship can only occur in the context of the migrant labor shortage in southern China since 2004. The fundamental concern is the vagrant identity of this new generation of young migrant workers who are uncomfortable with their farmer status in their hometown yet are excluded from the host society in which they have been working.

Concerning the issue of migrant worker identity, Chap. 10, "Mobile Communication and the Issue of Identity: An Exploratory Study of the Uses of the Camera Phone Among Migrant Workers in Southern China," attempts to explore the issue from a new perspective. Pui-lam Law's research on the uses of the camera phone among the migrant workers has unraveled an emerging phenomenon in which the workers consciously or unconsciously hide their worker identity through their visual representation. The use of ICTs, particularly the diffusion of the camera phone and the social networking system, has contributed to the possibility of the reconstruction of a worker's identity. If the thesis stands, it seems that the traditional labor capital framework fails to explain the social situation of the workers and that a theory of new connectivities should be developed to further capture the issue of identity among the migrant workers.

New Network and New Identification

In traditional China, the social order is achieved within the hierarchical social and political system. Capable agents or actors are directed to learn Confucianism, are channeled into officialdom, absorbed by the system through being rewarded with privileges, and consumed in the zero-sum game. It is an effective and resilient system for maintaining the established social order. Even in post-imperial China, its essential features have never really seemed to have disappeared. However, the penetration of the new connectivities seems to be undermining all that. In other words, various opportunities are emerging to the users who enjoy interacting, sharing information, maintaining traditions and/or embarking on new adventures.

In Chap. 11, "Beyond Privileges: New Media and the Issues of Glocalization in China," Boxu Yang maintains that the diffusion of the ICTs in China has begun to challenge deep-rooted traditions and will probably contribute to new social change in China. He cites the examples of the popular internet forums in China such as *Maoyan Kanren* (Looking at Human Beings from a Cat's Eye) and *Tianya Zatan* (Different Views in the Remotest Place on Earth) and illustrates that the discussions there either criticize traditional Chinese culture or unravel the problems of structures and systems or institutions in China. In this respect, the interactions of the virtual do indeed pose a challenge to the actual world. Chinese society has been experiencing new negotiation, and there is a possibility that the traditional social network may be gradually transformed into a new kind of social network.

Chung-tai Cheng's discussion of the *Super Girl* TV show and mobile voting in "The Principled Machine: A Sociopolitical Inquiry of Mobile Voting in Chinese Society" evidently support Yang's thesis. Cheng in his chapter tries to demonstrate that the combination of the old and new media technologies has induced the Chinese to take part in a new social participatory experience. Despite tele-voting being considered merely a kind of commercial entertainment, the Chinese government became alarmed and tried to contain this tele-participation. This case highlights the unpredictability of the diffusion speed and dimensions of ICTs and shows they have already started shaking the traditional hierarchical social and political system. The negotiation between the virtual and actual is inexorable once it has been ignited.

In addition to the challenge to the deep-rooted traditional social order, the social networking system (SNS) introduced by the Internet has provided a platform for new social networks in China. The case study in "The Use of ICT Products and "White-Collarization" of White-Collar Workers: An Everyday-Life Perspective" reveals the power of the SNS in the formation of the white collar network and the process of white-collarization. White-collarization is a process by which white collar aspirants actively or passively change their ways of life until they believe that they have actually become a real member of the white-collar group. By analyzing the uses of the Windows Live Messenger system (MSN), Yang and Li in their chapter find that the interaction of the ICT products and these workers endows them with fresh and different meanings, which strengthens their identification with their own groups, forming a specific system of membership identification, and solidifying the white-collar network, both in the virtual and actual world. The discussion of the new connectivities and the white-collar network offers a new perspective on social stratification theory.

While MSN tends to be utilized by the middle class Chinese, QQ, one of the largest and oldest SNS in China, is very popular among students and migrant workers. Larissa Hjorth and Michael Arnold, in "Home and Away: A Case Study of Students and Social Media in Shanghai," explores the role of QQ—and SNS more generally—through a case study of university students studying in Shanghai. As Hjorth and Arnold posit in their chapter, QQ has become a rite of passage for the migrating youth leaving home to study in another city or country. Through QQ, as a micro-lens for emerging technocultures and attendant forms of mobility, Hjorth and Arnold maintain that we can gain insight into new forms of media literacy. The migrating youth, a new generation of media literates, is using SNS to not only keep in contact with family and friends at home but they also often transfer their new media knowledge to their parents and grandparents. Hjorth and Arnold further argue that these new connectivities have also marked a new pathway of lifestyle cultures in China.

New Connectivities and Chinese Social Context

Part V focuses on the negotiation of the virtual connection, actual interaction, and the local culture in China. In "Toward a New "Electrical World": Is There a Chinese Technological Sublime?," Matteo Tarantino puts forward a possible framework for analyzing the cultural framing of the technologies of new connectivities such as the computer in China. The overall idea of this chapter is to identify the kind of emotional investment a culture operates in a technology by analyzing social discourse interplay: the object being transformed, the nature of the transformation, and the projected final outcome. In the Western case, this framing appears to be inherited from Christian metaphysics, as the technology is presented as offering the individual a way to transcend earthly limitations. The Chinese case appears different. Drawing from literary materials of various origin, a broad framework is identified in which technology is discursivized as a way for the nation to transcend a historical condition of disadvantage rooted in the "Century of Humiliation." While the "liberating" aspects of new connectivities technology are not negated, they are directed through a collective subject instead of an individual one. This paradigm gives Chinese politics a steady grip with which to regulate technology-related areas, while at the same time undermining Western assumptions about any inherent "democratizing" nature of technology: the layer of cultural meaning must be taken into consideration when reckoning the social effect of the new connectivities. Tarantino further argues that the paradigm elaborated by elites is negotiated by Chinese subjects in ways that are strongly entangled in the deep structures of Chinese society. The great national rebirth through technology envisioned by philosophers, politicians, and novelists is reduced to a slight improvement in the quality of life, without any trace of sublime images or feelings. Moreover, the "mass discourse" appears to be pervaded by a powerful dystopian fear of social disaggregation related to technology, which gives shape to almost-folkloristic demonic figures such as the "computer ghosts."

Culture indeed plays an important role in shaping the use of technologies. When we observe the Chinese ways of using the mobile phone in public places, perhaps we may agree with the studies that suggest that face-to-face interactions are often disrupted by mobile users engaging in private communications with absent-present others, leading to the possible conflation of the traditional separation of public space and private life. This seems to lend support to the argument of techno-determinism. But when we delve deeper into the issue of mobile phone use in China, we become aware of the working of culture on the new connectivities. Chinese mobile users, similar to users in the other countries, often transgress the norms of public communication. Wai-chi Chu explores this, in "ICT Use with Chinese Characteristics," and examines the way in which the Chinese concept of the relational shelf shapes mobile communication. Chu also maintains that some cultural elements, such as *mianzi* (face), have a significant effect on the use of the mobile among the migrant workers. Culture is one of the dominant factors influencing the adoption of the new connectivities in China.

Jinqiu Zhao shares a similar perspective in her study of the adoption of the Internet in rural areas in China. Despite the fact that China has the highest number of Internet users in the world, the vast population in China's hinterland has not been able to access this new technology, leading to a digital divide in the communication sphere. However, many believe that the Internet represents an opportunity for rural communities to improve their economic conditions and advance education. A number of experimental projects have been established to test this hypothesis. Zhao, in "ICTD: Internet Adoption and Usage Amongst Rural Users in China," studies the effects of Internet use on various aspects of rural life in China. She maintains that the introduction of the Internet in rural areas has not resulted in fundamental changes to the modes of production, the organization of work, and the norms of consumption.

However, it has had some positive effects: the Internet has increased farmers' access to advanced agricultural technologies and enhanced communication of farming innovations, and these developments have contributed to agricultural productivity. Zhao also maintains that the Internet has served an educational purpose by improving the level of technological awareness and readiness. Although the new connectivities have led to some improvement of rural social lives, Zhao warns us not to overestimate their effect on rural problems. She concludes that the diffusion, adoption, and uses of the new connectivities are all shaped by the social context.

Part I
The Internet Communication and the Issue of Civil Society

Chapter 1
NPOs in China: Some Issues Concerning Internet Communication

Boxu Yang

This chapter explores whether NPOs are able to benefit from new forms of media communication. While the implementation and dissemination of media technologies have been the catalysts of social, economic, political, and cultural changes in the West, this has not always been the case in China. For instance, the printing press technology invented in China nearly 1,000 years ago was not associated with any important changes in Chinese society. This clearly shows that a medium will remain merely a hardware unless it is discovered and employed by agents who use it to their advantage. It is, therefore, important to identify these agents when a new medium is introduced in order to gauge its impact. NPOs have become increasingly active in economic reforms as a result of the process of globalization in China (Li and Yuan 2005). It is, therefore, interesting to explore whether NPOs take advantage of the possibilities afforded by Internet communication.

1 NPOs and the Attributes of the Internet

The role the media has played in the past is relatively easy to gauge because the ways agents have used various media are matters of historical fact. However, both NPOs and the Internet are still in the early stages of development in China. The relationship between an agent and a form of media will be fruitful only if the agent's status matches the medium's attributes. If this is not the case, the relationship may be strained. Each medium has its own unique attributes. For instance, mass media favor elite or authorities because they are broadcasting in nature, while the Internet seems inherently grassroots-oriented because it is interactive and open to all.

B. Yang (✉)
School of Journalism and Communication, Peking University, Beijing, China
e-mail: byang@pku.edu.cn

P.-l. Law (ed.), *New Connectivities in China: Virtual, Actual and Local Interactions*,
DOI 10.1007/978-94-007-3910-9_1,

The mass media in China are tightly controlled by the government. They serve the government's political purposes and business's commercial purposes. The elite who are identified with these establishments are also able to voice their concerns via the mass media. All other agents are restrained and assumed to be members of the audience. China's NPOs, like those in other countries, are organized on a voluntary and nonprofit basis precisely because of their ideological differences with the political and business establishments. Thus, they are at a disadvantage when it comes to using the mass media in China. The implementation and dissemination of the Internet, however, is changing communication patterns. The latest report indicates that there are 162 million Internet users in China (CNNIC 2007).

In order to discuss the role of the Internet, we need to understand its characteristics and their implications. Such an understanding may be achieved by exploring how users' actions are facilitated, restrained, or sanctioned by the medium. The Internet is interactive and open to all. Usually, users are able to maintain their anonymity. Moreover, the Internet is easy to access, has unlimited space, and promotes the sharing of information. These characteristics compel the elite to interact with others on an equal footing if they want to be effective users. The elite are always in the minority; in the virtual world, members of the elite must constantly earn their respect based upon the quality of their input. They cannot simply rely on the status they enjoy in the "real world."

It is hard to classify NPOs as either elite groups or grassroots organizations for they differ significantly in terms of their missions and memberships. For instance, if we just consider the environmental nongovernmental organizations (ENGOs) in China, there are at least seven types (Yang 2005). The concepts of civil society and NPOs are relatively new in China. Nevertheless, the number of NPOs has been increasing dramatically since 1978. Wang (2004) maintains that we are witnessing a Chinese NPO revolution. In 2006, there were over 346,000 NPOs in China (MCA 2007), and 90% had been established since the economic reform (Li and Yuan 2005). Approximately 46% are private and nonprofit (MCA 2007). They often have small memberships and few financial resources (Li and Yuan 2005). Generally, they possess the following characteristics (Zhu 2004):

- A relatively higher autonomy
- An informal structure (often not registered with the government)
- Horizontal and networked relationships
- Volunteer membership and participation
- An ability to have a visible impact on the communities or groups they serve, although they lack the attention of the general public
- An organization consisting entirely of volunteers

Clearly, the attributes of the Internet—accessibility, interactivity, anonymity, unlimited space, and networking potential—suit the needs of grassroots NPOs. However, as the following discussion demonstrates, the relationship between the Internet and Chinese NPOs is complicated. It seems that it is concerned Internet users, and not NPOs, who are able to take full advantage of Internet communication in their efforts to construct a civil society in China.

2 Internet Users and Internet Communication in China

According to Frumkin (2002), the nonprofit and voluntary sector has four roles: service delivery, social entrepreneurship, the encouragement of civic and political engagement, and the promotion of values and faith. While the Internet can facilitate all four roles, it is particularly suited to the role of encouraging civic and political engagement. This involves mobilizing citizens for political action, advocating for causes, and building social capital within communities (Frumkin 2002). Obviously, one cannot rely on the government, the elite who are identified with the government, or other who benefit from the status quo to initiate the kind of political changes that NPOs endorse, particularly given China's political context. These elite groups are drawn to the mass media, which, as gatekeepers, they are able to manipulate and exploit to their advantage. To them, the general population is made up of audience members, not agents. However, new media such as the Internet are changing this image in an unprecedented way.

Concerned Chinese Internet users have been using the medium to engage in civic and political debate. While its universal accessibility provides a huge forum, its anonymity reduces the risks associated with free communication. These conditions are particularly important to the Chinese as far as civic and political engagement is concerned. There was no such public forum in China before the Internet. Virtually, all the platforms were monopolized by the authorities or the elite identified with the authorities. The Chinese who want to participate in civic and political affairs have suddenly found an opportunity to do so through the Internet. The fact that the Internet is open to all and allows users to maintain their anonymity seriously challenges those who feel they should be privileged in their Internet communication because of their status. The technological features of the Internet force all users to be equal when they communicate with each other on platforms such as the Bulletin Board System (BBS).[1]

As a result, the issues discussed on the Internet are more likely to be decided by the majority of the users than by the gatekeepers who, in China, are usually the mouthpieces of the authorities.[2] Indeed, the BBS is vital to Chinese Internet users who are involved in civic and political issues because there is no equivalent platform offered elsewhere.

The new media are grassroots-oriented. Millions of Chinese men and women (particularly young adults) who do not have the resources to voice their opinions via the mass media or solicit the information they require have found opportunities to do so on the Internet. According to the CNNIC report of July 2007 (CNNIC 2007b), about 66% of Chinese Internet users receive a monthly salary of about 1,500 *yuan* (roughly 150 Euro); more than 56% have a high-school education or less; more than 51% are younger than 25[3]; and more than 60% are unmarried. While such statistics

---

[1] BBS is dedicated to the sharing or exchange of messages or other files. A BBS may be accessible from a dial-up modem, Telnet, or the Internet. Most bulletin boards serve specific interest groups.

[2] It should be pointed out that Internet communication is still censored to some degree in China.

[3] More than 70% of users are 30 or younger.

cannot be used to analyze specific Internet communicative actions or patterns, they do indicate the socioeconomic status of the majority of users.

It is clear that millions of users employ the Internet for BBS purposes. According to the CNNIC report of January 2007 (CNNIC 2007a), almost 40% of Chinese Internet users participate in the BBS forums.[4] They are free to discuss the topics of their choice on the BBS. These topics include "corruption," "government policy," "law and order," "sex," "military action," "foreign affairs," "social relations," among others. It does not matter if the views are liberal or conservative. The most influential forums on BBSs are able to provide channels for any viewpoint. However, using an authoritative or elitist approach meets with disapproval. Such a tone usually causes embarrassment and suspicion. After all, Internet communication is supposed to be a form of sharing. Internet users are not interested in being lectured, at least not on the Internet. The Chinese elite and various authorities have enjoyed the one-way communication afforded by mass media where they could set the agenda. This is particularly true in the area of civil and political engagement.

Chinese Internet users seem to have lost confidence not only in authority but also in organizations in general. While they are politically engaged and very responsive to the idea of civil society, they do not associate their Internet activities with the work of NPOs. Rather, they take a "do-it-yourself" approach to civil and political engagement. The essence of this model is interactive communication. This model does not recognize the authority of an institution or individual status. As a popular posting on the major Chinese websites states, "Just as the corruption we are experiencing cannot lead to a new [social and political] system, going down on our knees and begging will not yield a civil society either. Do not dream that apple pies will drop from the sky. Do It Yourself! A sound civil society depends upon the DIY efforts of hundreds and millions of ordinary human beings like you and me" (Cai 2007).

Such DIY messages were triggered by an incident of severe environmental pollution in rural China (Cai 2007). The incident was first reported by a well-known Chinese newspaper, *China Youth Daily*. The story was then posted on *Sina*, one of the major Chinese websites, on July 4, 2007 (Li 2007). According to the newspaper, for the past ten years, the physical health of the residents of a village called Baojiagou in Anhui province has been seriously threatened by a polluted river. The river, which the villagers depend on for their farming, has been polluted by small nearby factories. The residents have repeatedly begged the local and provincial governments for assistance. On May 26, 2007, some of the villagers knelt before a high-ranking official of the state inspection team from Beijing when the team reached the village on their tour. The villagers believed that the authority from Beijing represented their only chance to solve their environmental problems. While the environmental NPOs have been doing work in China, they often hesitate to participate in civil and political matters. In fact, the Chinese NPOs openly acknowledge that they are cooperative with the government (Wexler et al. 2006).

---

[4] It is interesting to note that the latest report released by CNNIC (2007) has no statistics on BBS usage.

Chinese NPOs seem to have discarded the role of advocate in order to focus on their own initiatives (Wexler et al. 2006). However, Chinese Internet users have taken up the mission of advocacy in a nonorganizational and nonstructural way. Users post their ideas or opinions on the BBS or their own blog and receive responses from others who share their concerns. This was exemplified in 2006 when a widely circulated and hotly debated posting appeared on Chinese websites questioning the term "common people," which Chinese authorities often employ. The authors of the posting insisted that the Chinese people should be referred to as "citizens," a term that does not have the negative connotations associated with "common people." In fact, the title of posting was "Call Me Citizen, Not Common People."[5]

The significance of this posting lies not only in its mobilizing function but also in its capacity for enlightenment: the posting clarifies some important concepts. For instance, the posting points out that "common people" is the term employed when an emperor or dictator refers to his subjects, while "citizen" is associated with rights, freedom, and democracy. Thanks to the virtual public spaces on BBSs that enable such discussions, users are willing and able to question the slogans promoted by the authorities and to learn their true implications.[6] For the first time in Chinese history, the authorities have lost the privilege of being the only purveyors of "truth" due to the "virtual public space." Chinese Internet users can function as independent agents and define the social, political, and economic reality based upon their own life experiences and knowledge. They are no longer interested in being "represented" or "mobilized" by the authorities or the establishment.[7] They want to empower themselves through sharing information and knowledge with fellow users. In so doing, they advocate for causes and build social capital within the virtual communities.

3 The NPOs and Internet Communication in China

Mobilizing citizens for politic action, advocating for causes, and building social capital are necessary elements in the creation of civil society in China. Considering the long history of China's autocracy, the passivity of the "common people" is habitual; they are used to a world where "masters" dominate. One of the major difficulties for the NPOs operating in China is of the lack of civil awareness among the public. The NPOs, particularly those that are small and grassroots-oriented, often

[5] The author of the posting used the nickname "light as heavy" to preserve anonymity. From http://www.tianya.cn/publicforum/Content/free/1/450349.shtml (in Chinese). Accessed August 2, 2007.

[6] The concepts they questioned included GDP, development, national conditions, and tax.

[7] Indeed, this has been a recurring theme in the BBS and blogs. I typed "I don't want to be represented" in Chinese into Baidu, a well-known search engine in China, and 16,900 web pages were found. From http://www.baidu.com/s?wd=%B2%BB%CF%EB%B1%BB%B4%FA%B1%ED&cl=3. Accessed August 4, 2007.

complain that the public does not have the knowledge or the willingness necessary to create a civil society, which is due, at least in part, to China's long history of autocracy. Moreover, the current Chinese regime exerts a strong influence on the volunteer sector (Huang 2004). While it may be true that the public is passive in its pursuit of civil society in China, it is not true that they lack the knowledge or willingness to be agents of civil and political change in the virtual world.

Civil and political engagement demands communication. However, interactive communication is more suited to the process than mass communication. This is not only because the mass media are tightly controlled by the government but also because the "mass" needs to learn and practise rational and polite means of discourse when discussing civil and political affairs. After all, they have never been allowed to do so before. Now, they can go to the Internet. Currently, there are over one million web pages containing "NPO" in Chinese. All demonstrate an awareness of the importance of NPOs in Chinese society. Approximately 168,000 of these pages are blogs[8] of people who are concerned with the role of NPOs in Chinese society. These bloggers represent a pool of talent that could be recruited by the NPOs. However, for the most part, Chinese NPOs appear to be uninterested in these Internet users.

Chinese NPOs rarely post messages on the Internet in order to mobilize citizens for civil and political engagement. They seem to be indifferent to whether Internet users support their daily operations in the field. For instance, we all know that environmental pollution has become a serious problem in China. However, we rarely see firsthand information on the situation from Chinese environmental NPOs on the Internet. Instead, NPOs rely on the mass media for communication and on meetings with government officials in order to voice their concerns or get resources (Li and Yuan 2005). In their efforts to promote civil society, they ignore the Chinese people themselves. As a result, Chinese NPOs have become increasingly bureaucratic and elitist. Their leadership style has begun to imitate that of Chinese officials (Wu 2006; Jia 2007). This attitude is not suited to Internet communication, so they do not adopt it.

The Chinese government, elite, and mass media are often harshly criticized by the Internet users on the BBS. In return, the authorities openly label some Internet users as "violent mobs" and attack them through the mass media they controlled (Deng and Wang 2007; Tong 2006).[9] It is true that some of the discussions on the BBS forums are not rational political arguments by Western standards. In fact, the discussions on the BBS are often highly charged emotionally. This is understandable since the vast majority of Internet users have never really had the

[8] These numbers are obtained through Baidu, a Chinese search engine, by typing the keyword "NPO" first. And then the keyword "blog" was searched within the identified NPO-related web pages. From http://www.baidu.com/. Accessed August 5, 2007.

[9] The People website is the main Internet platform of Chinese Communist Party. As of August 13, 2007, it carried 276 postings criticizing "Internet violent mobs." From http://www.people.com.cn/. Accessed August 13, 2007.

opportunity to discuss public matters openly in China. The new media have empowered the "grassroots" and given ordinary people the chance do things that they were never allowed to do in Chinese history. Above all, the Internet users have become agents in the quest for knowledge and freedom of expression. However, these agents are not encouraged by the NPOs, even though both parties share many of the same goals.

It is odd that the Chinese NPOs refuse to take advantage of the opportunities that the Internet provides. It seems that the Chinese NPOs do not fit the Western definition of NPOs. Their tendency to take a hierarchical and elitist approach prevents them from working with concerned Internet users. The Internet is a forum where users interact, not where organizations lecture. By its very nature, the Internet does not allow the Chinese NPOs to assume the role of "parents," as Chinese officials often do. In order to use the Internet effectively, representatives of NPOs would need to function as agents who are themselves members of the "common people" and to share their information (rather than exert their authority) to empower and mobilize their fellow citizens.

4 Concluding Comments

The characteristics of the Internet encourage freedom and equality of interactions. The implementation and dissemination of the Internet has not only enabled users to transform themselves from passive audience members into active users in the virtual world but also robbed the government and establishment of their position as the sole purveyors of the "truth." This has been accomplished by the new media in general and the Internet in particular. Internet users are able to act in unprecedented ways—as agents in Chinese history. Although the transformation from audience members to agents may not automatically lead to the emergence of the kind of action required for a civil society, it is very encouraging. The Internet provides a "public space" where activities such as political mobilization, advocacy, and social capital building are possible.

While concerns and opinions are often expressed in virtual forums such as the BBS, rational discussion and mutual respect are often lacking. Since Chinese people have no experience in public debate, it is quite understandable that they would not be immediately adept. The important point is that the Internet empowers users who are willing to share their knowledge and teaches them effective means of discourse. Chinese NPOs shy away from this mode of civil and political engagement in order to avoid direct conflict with the government and business establishment in both the real and the virtual world. Their elitist and instrumental approach to civil society often puts them in embarrassing and difficult situations. As a consequence of this approach, they have not engaged the public's attention, and they have not recruited a number of highly committed individuals. It is difficult to predict the sustainability of Chinese NPOs. However, it is clear that their approach to civil society would have to be radically reformed in order to take full advantage of the possibilities afforded by Internet communication. The fact is that Chinese Internet users are not paying attention to the NPOs, although they share the goal of NPOs—to create a civil society.

A civil society cannot be achieved without agents who actively participate in the civic and political issues. It seems that the Chinese NPOs have no choice but to reposition themselves as a grassroots movement if they want to fit into the new media environment. A good first step would be to identify the Internet users who have shown their commitment to the idea of civil society in their blogs or postings on the BBS. NPOs must recognize and appreciate the attributes of the Internet as well as the contribution of Chinese Internet users. If they do not, they are unlikely to become effective agents for a civil society in China.

References

Cai, Z. (2007). DIY a civil society. From http://news.sina.com.cn/pl/2007–07–27/102413539768.shtml/. Accessed 31 July 2007 (in Chinese).

CNNIC. (2007a, January). Report of China Internet development statistics. China Internet Network Information Center. From http://www.cnnic.net.cn/uploadfiles/doc/2007/2/13/95848.doc. Accessed 10 Aug 2007 (in Chinese).

CNNIC. (2007b, July). Report of China Internet development statistics. China Internet Network Information Center. From http://www.cnnic.net.cn/uploadfiles/doc/2007/7/18/113843.doc. Accessed 26 Nov 2007 (in Chinese).

Deng, X., & Wang, S. (2007, August 10). See no to the abusive opinion of mobs on the Internet. *People's Daily*, p. 16 (in Chinese).

Frumkin, P. (2002). *On being nonprofit: a conceptual and policy primer*. Cambridge, MA: Harvard University Press.

Huang, H. (2004). Strategy management for NPO development in China. *Volunteer Service Journal, 6*(4), 9–16 (in Chinese).

Jia, P. (2007, August). NPOs become profit organizations: The Chinese organizations full of bureaucratic airs. *China Economic Time*. From http://www.chinaelections.org/NewsInfo.asp?NewsID=114216/. Accessed 10 Aug 2007 (in Chinese).

Li, L. (2007, July). Lost confidence on local government, Anhui villagers kneeled before the officials of state environmental protection administration of China and begged for solving the problem of pollution: a special report. *China Youth Daily*. From http://news.sina.com.cn/c/2007–07–04/054013369531.shtml/. Accessed 31 July 2007 (in Chinese).

Li, J., & Yuan, R. (2005, October). *The development of civil society in transitional China: an analysis from the perspective of local social organizations*. Paper presented at The international conference on civil society development in transitional China, Beijing (in Chinese).

MCA (Ministry of Civil Affairs). (2007). *Statistics on the development of civil affairs 2006*. From http://www.mca.gov.cn/. Accessed 9 July 2007 (in Chinese).

Tong, D. (2006, April). The Internet panic syndrome of Chinese elite intellectuals. *Oriental Morning Post*. From http://tech.sina.com.cn/i/2006-04-25/0916916609.shtml/. Accessed 13 Aug 2007 (in Chinese).

Wang, S. (2004). The Chinese NPO revolution: an overall mapping of the Chinese NPOs. *Zhejiang Academics Journal, 6*, 71–77. From http://www.ngocn.org/Article/ShowArticle.asp?ArticleID=1212/. Accessed 9 Aug 2007 (in Chinese).

Wexler, R., Xu, Y., & Young, N. (2006, September). NGO advocacy in China: a special report. *China development brief*. From http://www.chinadevelopmentbrief.com/node/749. Accessed 12 Aug 2007.

Wu, B. (2006). Constructing Chinese civil society network from the bottom. Center for Huanan NPO Studies, Zhongshan University. From http://www.ngocn.org/Article/ShowArticle.asp?ArticleID=1212/. Accessed 9 Aug 2007 (in Chinese).

Yang, G. (2005). Environmental NGOs and institutional dynamics in China. *China Quarterly, 181*(3), 46–66.

Zhu, J. (2004). Grassroots NGO and the growth of Chinese civic society. *Open Times*, *174*(6). From http://www.opentimes.cn/to/200406.htm/. Accessed 11 July 2007 (in Chinese).

Chapter 2
Dot the "I's" and Cross the "T's": A Sociological Interpretation of Chinese Cyberspace and the Qianyuan Grace Wang Incident

Chung-tai Cheng

1 Introduction

With the prevalence of Internet use in China, some scholars believe that Chinese people now have more opportunities to express and discuss their own views on social matters, which in turn may contribute to the view of the Internet, in a Habermasian sense, as a public sphere which may enhance and promote democracy in China (Lowell and Guoli 2006; Wu 2007; Zheng 2008). In contrast, more and more ordinary Chinese people worry that unrestricted use of the Internet may lead to a loss of control or threaten personal privacy. Both perspectives share a similar assumption that the social influences of the Internet can be positive and/or negative, depending on the choices people make. In other words, the Internet is an instrumental means of maintaining and fulfilling human needs.

This idea, however, may oversimplify the interaction between the Internet and very specific sociocultural contexts. Though the innovation of the Internet was intended for communication between scientists and for military use, it is now ubiquitous in many societies. This means the Internet blends tightly with local cultures, rather than merely being a simple and neutral tool. If this is so, the central question is how the Internet intertwines both actively and passively with the society in which it is situated. On the one hand, the Internet has its own characteristics that may reframe or reconstruct people's understanding of and communication about events. Because of the limitations of hyperlinks, the possibility of understanding and commenting on events may be limited to the layer of emotional preconception (Dreyfus 2001). On the other hand, even though the Internet tends to be regarded as a handy instrument in Chinese cyberspace, the hidden beliefs behind this kind of means-end

C.-t. Cheng (✉)
Department of Applied Social Sciences, The Hong Kong Polytechnic University, Hong Kong, China
e-mail: sscct@inet.polyu.edu.hk

P.-l. Law (ed.), *New Connectivities in China: Virtual, Actual and Local Interactions*,
DOI 10.1007/978-94-007-3910-9_2,

relationship are important if we wish to investigate the normative implications of the Internet toward Chinese society (Feenberg 1999).

Drawing on a case study of the "Qianyuan Grace Wang incident" in China during the spring of 2008, the first section of this chapter discusses how the Internet conditions Chinese people's understanding of the event, in particular of Wang's intentions and her "T" hand signal. It then tries to explore how Chinese netizens interpret and comment on Wang's behavior based upon a thin and shallow understanding of the event. Because of the leveling off of real Chinese relationships in cyberspace, the author proposes an answer to the question why Chinese netizens were likely to be involved in the event and given the rights and duty to discipline and punish Wang, whether in the real world or in a cyber one. This chapter concludes that the Internet cannot be simply understood as an objective means to achieve human needs because the representation of Chinese cyberspace is embodied and co-constructed by Chinese indigenous cultures.[1]

2 A Brief Description of the Grace Wang Incident

In April 2008, while the Beijing Olympic torch was being carried through San Francisco, USA, Grace Wang ran into a group of American students and used blue body paint to write "Free Tibet, Save Tibet" on the bare back of one of the organizers. Since she had both pro-Tibet and pro-China friends, she tried to act as a mediator and get the two groups to talk at a meeting on her university campus. Before long, a video of the Chinese freshman at Duke University, seen standing between pro-Tibet activists and Chinese counter protestors, was posted on the Internet along with several pictures. In one photo, Wang was walking toward a phalanx of Chinese flags and banners with her arms overhead in a "T" hand signal. Another photo also appeared on an Internet forum for Chinese students with a photo of Wang and the words "traitor to the motherland."

As soon as the images appeared online, Wang not only was accused of being a traitor, but she also became one of the targets of the "human flesh search engine" in China. Her personal details, including her Chinese name, national identification card number, and high school and contact information, were posted on the Internet, along with directions to her parents' apartment in Qingdao, a port city in Shangdong Province. The story of the young woman who was said to have betrayed the nation and the people by taking the side of pro-Tibetan independence groups spread through China's most popular websites, at each stop generating a number of angry posts and derogatory comments rooted in nationalist sentiments. Even worse,

[1] The data in this chapter were obtained through *Baidu*, a famous Chinese search engine, by typing the keyword "Qianyuan Qianyuan Wang Incident" (in Chinese), resulting in 56,500 web pages being found. From http://www.baidu.com/s?wd=%CD%F5%C7%A7%D4%B4%CA%C2%BC%FE&oq=%CD%F5%C7%A7&f=3&rsp=1. Accessed January 15, 2009.

Wang's family home in China was daubed in excrement and "Kill the Whole Family" was painted in big red letters in the hallway of her apartment building. Eventually, her parents had to temporarily move away.

3 The Will to the Internet

According to some net enthusiasts, the age of the open Internet and citizen journalism enables society to achieve a new level of civilization. As Dan Gillmor, one of the foremost proponents of this vision, puts it, "People become more engaged with the events around them, and especially to the extent that they become journalistic activists, they will be making better citizens of themselves. Tomorrow needs them" (Gillmor 2006: XVIII). Apparently, this idea reveals a strong belief that the Internet cannot only provide a communicative platform for ordinary people to comment on and discuss social matters, but it can also help build a better and democratic future. Chinese netizens are able not only to gather information online but also to respond to, comment on, and become engaged in instantaneous discussions on online forums. Especially, the Internet may offer a greater hope for improving Chinese society since it can repair and mediate the interstices between the failure of laws and the loss of morality in contemporary China (The People 2008).

However, the social implication of the Grace Wang incident is not only that it shows "the medium is the message" (McLuhan 1997) but that it also indicates that the Internet is far more than the message. Though some commentators concluded that the actions of the Chinese Internet lynch mob in the Grace Wang incident were just an example of Chinese nationalism and blind patriotism (Tang 2008), this chapter wants to suggest that before exploring the power of the Internet to affect people's understanding and interpretation of events, it is not appropriate to claim Wang's experience is only the reflection of traditional Chinese practices, a reminder of the bitter experience of the Cultural Revolution between 1966 and 1976. Rather than simply showing how the Internet became a cyber execution ground in the hands of angry Chinese youth or *Fen Qing*, the Grace Wang incident raises some questions. Why was the "T" hand signal interpreted as a "T" for "Free Tibet" rather than for "time out" in Chinese cyberspace? Also, why were Wang's actions attributed to her desire for a green card[2] and personal promotion, rather than promoting freedom of speech and fighting for a movement? In short, how do netizens understand and interpret the event through this kind of Internet-mediated communication?

First, it is quite clear that the Internet changes our form of communication. The popularization of the Internet in China not only promotes freedom of speech but also freedom of the press. Chinese people can be citizen journalists as long as they have a digital camera and are connected to the network. It means everyone can say

[2] Green card is the nickname of the US permanent resident card because of what it used to be colored. It is an identification document, showing that someone who is not a citizen of the USA has permission to live and work there.

"I have discovered, I have seen, and I know" in cyberspace, whether it really happens or not. The implication of this phenomenon is that, though the Internet offers facts, the gap between true and false and real and imaginary is narrowing and becoming blurred (Baudrillard 1994). More importantly, the way the Internet works is based on each element of disorganized information linking up with many other elements by hyperlinks. Since there is no overall authority governing accuracy and no agreements binding this catalog system that may constrain the linker's associations, everything can be linked to everything else without regard for purpose or meaning (Dreyfus 2001). Thus, the classification of information in cyberspace may not only be impossible but also meaningless. With this difficulty in locating specific information on the Internet, the critical issue here is how netizens associate and confirm what kind of information is relevant to others and to themselves.

Second, with the advancement of communication technology, Chinese people can enjoy another form of presentation and self-identification. Chinese netizens can express views and opinions in a comparatively convenient and compact way, in which the whole process of presentation only requires the pressing of a button and a wait of a few seconds. Since the massive distribution of piecemeal information on the Internet is making every sort of information immediately available to everyone, Chinese people are encouraged to develop and express an opinion about everything only if they have a connected computer. The democratization of the public sphere is henceforth attributed to this form of communication, but, learning from the Grace Wang incident, the public sphere itself may also be reframed and reconstructed by Chinese interpretations of the Internet. During the incident, each person took sides, agreeing or disagreeing with Wang's action within seconds as if they were part of the event. Even if they did not participate in the process of online discussion, those who remained silent may still have been part of the cyber hunt and cyber bullying of Grace Wang. Therefore, apart from merely viewing the Internet as a means of recapturing the moral and political virtues of the public sphere, it is important for us to consider how the social meaning of the Internet is co-constructed by its own power and indigenous Chinese cultures.

4 Cyberspace as Chinese Family System

From the Grace Wang incident, we can see that the Internet is not only a medium that carries messages, but it can also create and implement messages at the same time and with the same means (Wu 2007). Though some Chinese netizens believed they could objectively make use of the Internet to search for or find out the "truth" of the Grace Wang incident through the collective power of Chinese cyberspace, and especially the Human Flesh Search Engine, they forgot that they were also engaged in the construction of the truth of the event. In other words, they not only can learn about distant things through the screen and the Internet, but they can also participate in an event as if they were right on the scene. Thus, how does an individual determine what elements are related to the event in the river of information?

More importantly, how do Chinese netizens make a comparatively fair judgment of the event through Internet-mediated communication? This chapter suggests that the Internet is guided by Chinese indigenous cultures, in which most Chinese netizens tend to justify Wang's behavior and their views by citing principles along with national and individual sentiments, whether in practice or in real life.

Having a public online communicative platform does not necessarily mean that Chinese society is more democratic than before. The power of Chinese deep structure may manipulate or even distort the condition of the possibility of establishing rational discussions on the Internet (Potter and Potter 1990). An interesting phenomenon in the Grace Wang incident is that most Chinese netizens' comments or discussions about the event were directed at Wang's family and those who were related to her. For instance, one netizen wondered how Wang's parents could produce such a barbarian and that they would choose to eat her like BBQ pork chops if she was their daughter; another netizen complained that Wang's class teacher should cut her throat from the shame of teaching such a female traitor (Tianya 2008). In short, a Chinese proverb can summarize these complaints about Wang's behavior: "To feed without teaching is the father's fault; to teach without severity is the teacher's laziness." In this point of view, the question how Chinese netizens interpreted their own roles and statuses in cyber communication becomes extremely important for us to understand the interaction between Chinese society and the Internet. Notably, when Chinese social relationships are understood as a "differentiated mode of association" (i.e., the closer the other is to the individual, the more intimate their relations will be (Fei 1992)), then what are the implications to Chinese society if the Internet levels off this kind of differentiated and graded social relationship?

Social interaction in face-to-face communication must be based on social norms; cyber interaction, however, may weaken the effectiveness of a normative foundation in a life world. Through body gestures, facial expressions, intonation, and the use of language in face-to-face communication, it is possible to obtain a comparatively thick description of the event under discussion. Accordingly, perceptions may be gradually shaped and reshaped through the process of reasoning and conversation. Though both face-to-face and mediated communication are not mutually exclusive, the key issue is how the differences between them may reframe or even reconstruct Chinese netizens' understanding on the Grace Wang incident. The Chinese notion of "self" is understood as being different from the rational, self-conscious, and autonomous Western concept of "self." It is a "relational self" which tends to be socially or psychologically dependent on others, for this "situation-centered" individual is tied closer to his world and his fellow men (Hsu 1981). It means the essence of Chinese relationships is not referred to as treating someone as an individual on equal terms, but rather according to relationships between other persons and oneself. To understand the dynamic of the interactions of Chinese relationships, one must participate in and learn from the experience of interacting and communicating with the others, both in practice and in real life, to learn how to behave in accordance with the time and the place, that is, if someone behaves socially and/or morally in an unacceptable manner, that person will be urged to modify their behavior. During this process, one's self-identity can easily be established by the notion of behaving with "*li*" (propriety). This is the reason why face-to-face communication in Chinese society is essential for maintaining social relationships, confirming one's own social identity, and strengthening social solidarity.

Unlike face-to-face communication, the normative foundation in the life world may be weakened and loosened in the cyber world. As there is a lack of real-life experience as a backdrop for the support of the separation of "front stage" and "back stage" (Goffman 1973), it may be impossible or even meaningless to take notice of one's own status and presentation on the Internet. Everyone in cyberspace is supposed to be an equal with the autonomy to determine their own position and gather every sort of information immediately, thereby producing a kind of desituated, detached spectator. Netizens can have an opinion about and comment on all kinds of public matters in cyberspace without needing any firsthand experience and without having or bearing any responsibility (Dreyfus 2001). While Chinese people share information and discuss issues on the Internet, they must experience a process of referencing. Netizens can only pick and choose the facts or information which is believed to belong to an event bit by bit. As Thompson maintained, "Mediated communication involved events that are distant in space and in time and that are refractory to the individuals who experience them… Mediated experience is not a continuous flow but rather a discontinuous sequence of experiences which have varying degrees of relevance to the self" (Thompson 1995: 230). However, this does not imply that Chinese netizens would treat each other individually on the Internet. While Chinese social relationships are understood as a reciprocal social network, establishing social networks of personal relation with the self at the center and decreasing closeness as one moves out, the Internet eats up all individuality's relativity and concreteness, and this is especially true of the Bulletin Board System (BBS). There is thus a paradox in Chinese cyberspace: on the one hand, Chinese netizens may hold indifferent attitudes toward others since they are strangers who do not have exact and concrete social relationships on the Internet (this is in keeping with the Chinese saying that "each one just sweeps the snow from his own doorsteps but doesn't bother about the frost on his neighbor's roof"). On the other hand, as social relationships are leveled off in cyberspace, the rings of the Chinese circle can be stretched out without limit. Chinese are supposed to be part of the events only if there are sensitive and flammable topics which contradict Chinese social norms, simply because the events are never regarded as public issues, but family business.

Since there is no concept of man as separate from men in Chinese culture (King 1994), the emergence of cyberspace has contributed to the peculiar situation that Chinese netizens can become either insiders or outsiders with each other at any moment in cyber communication. Thus, the following question is what the logic and the criteria of determining one's identity are in Chinese cyberspace, or how can Grace Wang be so easily excluded and labeled a traitor or *hangan*? And why did most of the Chinese netizens tacitly agree that the treatment of Wang was fair and appropriate? Before looking into these questions, we should bear in mind that the single English word "Chinese" cannot describe the complexity of Chinese identity, for it not only misses certain meanings but may cause confusion. In Chinese, there are many terms used to describe and reflect racial, ethnic, cultural, social, and national attributes of being a Chinese, both in spoken and written language, for example, *Zhonggouren*, *Zhonghua minzu*, *huaren*, *hanren*, *tangren*, *huaqiao*, and *hauyi*. “Chineseness” is not, therefore, a fixed and static concept, in which “Chinese people can conceptualize their own Chineseness in the peripheral situation, demonstrating the complex process whereby they are able to incorporate indigenous language and culture without losing their sense of having a Chinese identity” (Wu 1994: 148). In view of the confusing senses of ethnic and cultural identity among Chinese people, the Grace Wang incident reveals a more complicated but interesting phenomenon about Chineseness in the digital world. Of course most Chinese netizens would agree that Wang is undoubtedly a Chinese person in theory since she has not only grown up in China but also belongs to a great civilization. Afterward, the appearance of Wang’s images on the Internet forced Chinese netizens to try to understand and interpret the rationale behind Wang’s action, which did not accord with certain Chinese norms of conduct. Based upon a thin and shallow understanding of the Grace Wang incident through Internet-mediated communication, Chinese netizens discussed Wang’s behavior guided by indigenous principles. But like Potter and Potter maintained, “The free expression of emotion is not perceived as a threat to authority, and anger, *per se*...What is perceived as a threat to authority is inappropriate judgment, rather than inappropriate emotion” (Potter and Potter 1990: 185). From this point of view, there are only two possible ways that Chinese netizens can perceive Wang: one is as an insider to be reeducated; the other is to marginalize or even exclude Wang from the rings of the Chinese circle by de-Sinicizing her Chinese attributes: “her body movements, including her facial expression, body gesture, and the use of languages, were too westernized, exaggerating and aggressive as if she had already been brainwashed by the arrogant American” (Juliana 2008). Hence, Grace Wang is now neither Chinese nor non-Chinese, but rather a Chinese, who may have double identities in Chinese cyberspace: one is a “sociohistorical identity” which was given naturally when she was born; the other one is a “sociopolitical identity” which is culturally given and recognized by others. While her behavior could not and would not be accepted in Chinese society, her sociopolitical identity may not only be revoked, but she may also become an exception/exemplary model to others, with her efforts to explain and defend herself probably regarded as lies and excuses. However, we should notice that the importance of the Grace Wang incident is not only characterized by individual emotional expression but by the symbolic medium for the expression of social connection and the affirmation of social solidarity in the most fundamental terms the Chinese people know.

5 Conclusion

Though the Internet provides Chinese people with a new and compact means of communication and presentation, it does not mean that the Internet can be simply viewed as the retrieval of the Habermasian public sphere. From the Grace Wang incident, we learn that the interactions between the Internet and Chinese indigenous cultures are much more complicated than the ideal belief that the Internet can enhance and promote democracy in China. Any technology may be used in a multiplicity of ways, limited only by the individual, social, and cultural imagination of the people (Ihde 1993). Thus, the Internet is beyond either positive or negative influence since it is not only a technological product but also a social artifact.

Apart from benefiting social and economic development, the Internet offers Chinese people another form of communicative action. Even though communication between people is more convenient than before, Chinese netizens only have a thin and shallow understanding of the Grace Wang incident because of the absence of the real situation and the limitations of the hyperlinks that characterize Internet-mediated communication. Simultaneously, the differentiated and graded Chinese social relationships are leveled off in cyberspace, which may contribute to the possibility that Chinese netizens may regard each other as either insiders or outsiders at any moment in cyber communication. The implication is that the roles Chinese netizens perceive they have in cyberspace may determine the social meaning of the online public sphere in Chinese society, in which they may be indifferent to the topics that do not have a direct relationship with them or in which they may feel that they are obliged to be involved when events invoke national and cultural memory and sentiments. It matters not what views Chinese netizens hold; this chapter suggests that the presentation and the interpretation of the Internet are co-constructed by indigenous Chinese culture.

References

Baudrillard, J. (1994). *Simulacra and simulations*. Ann Arbor: The University of Michigan Press.

Dreyfus, H. L. (2001). *On the Internet*. London/New York: Routledge.

Feenberg, A. (1999). *Questioning technology*. London/New York: Routledge.

Fei, X. (1992). *From the soil: The foundations of Chinese society*. Berkeley: University of California Press.

Gillmor, D. (2006). *We the media: Grassroots journalism by the people, for the people*. Sebastopol: O'Reilly.

Goffman, E. (1973). *The presentation of self in everyday life*. Woodstock: Overlook Press.

Hsu, F. L. K. (1981). *Americans and Chinese: Passage to differences*. Honolulu: University of Hawaii Press.

Ihde, D. (1993). *Postphenomenology: Essays in the postmodern context*. Evanston: Northwestern University Press.

Juliana. (2008). *Wang Qianyuan shijian wanquan ban* [in English, The full version of Wang Qianyuan incident]. http://blog.bcchinese.net/juliana/archive/2008/04/17/132970.aspx. Accessed 3 May 2011.

King, A. Y. (1994). Kuan-hsi and network building: A sociological interpretation. In W. M. Tu (Ed.), *The living tree: The changing meaning of being Chinese today* (pp. 109–126). Stanford: Stanford University Press.

Lowell, D., & Guoli, L. (2006). *China's deep reform: Domestic politics in transition*. Lanham: Rowman & Littlefield Publishers.

McLuhan, M. (1997). *Understanding media: The extensions of man*. Cambridge, MA: The MIT Press.
Potter, S. H., & Potter, J. M. (1990). *China's peasants: The anthropology of revolution*. Cambridge: Cambridge University Press.
Tang, B. (2008). *Wang Qianyuan and the Internet lynch mob*. http://blog.foolsmountain.com/2008/05/07/wang-qianyuan-and-the-internet-lynch-mob/. Accessed 3 May 2011.
The People. (2008). *Renrou sousuo de zui yu fa* [in English, The guilt and punishment of the human flesh search]. http://media.people.com.cn/GB/40606/7764891.html. Accessed 3 May 2011.
Thompson, J. B. (1995). *The media and modernity*. Oxford: Polity.
Tianya. (2008). *Viewing net angry youth from Wang Qianyuan incident*. http://cache.tianya.cn/publicforum/content/funinfo/1/1128430.shtml. Accessed 18 Jan 2009.
Wu, D. Y. (1994). The construction of Chinese and non-Chinese identities. In W. M. Tu (Ed.), *The living tree: The changing meaning of being Chinese today* (pp. 148–166). Stanford: Stanford University Press.
Wu, X. (2007). *Chinese cyber nationalism: Evolution, characteristics, and implications*. Lanham: Lexington Books.
Zheng, Y. (2008). *Technological empowerment: The Internet, state, and society in China*. Stanford: Stanford University Press.

Chapter 3
Rage and Reflection: Chinese Nationalism Online Between Emotional Venting and Measured Opinion

David Kurt Herold

1 Introduction

In talking about nationalism, Farrer argues that it is a relatively recent phenomenon all across Asia where one result of the increasing globalisation of Asian countries is that the '*celebration of cultural hybridity and cosmopolitan chic in Asia's global cities goes hand in hand with the rise of cultural nationalism among Asian youth*' (2007: 109). Yuan argues a similar view, stating that in China during the 1990s

> a new brand of nationalism arose more organically among the people, as opposed to nationalism simply manipulated by the leadership [...], seeking time and again to check the country's progress toward openness and reform. (2007: 22)

In both instances, the writers assume that nationalistic feelings are an easily understandable defensive reaction of members of a traditional cultural sphere (Asia in general or China in particular) against the modernising influences of a globalising force of economic development based on the Euro-American model. This creates a questionable dichotomy between the 'good' forces of globalisation and their by-product, an enlightened (and 'Westernised') culture, and the 'bad' locals who are trying to hold on to their outdated traditions out of misguided feelings of national pride (see Yuan 2007: 27).

One result of this view – and the starting point for the discussion presented here – is how events on the Chinese Internet are portrayed in the non-Chinese media. Stories on the Chinese Internet published in non-Chinese media frequently focus on the nationalistic and 'anti-Western' slant of many websites and online communities (see e.g. Grossman 2006). In particular, during the run-up to the 2008 Olympic Games in Beijing, many non-Chinese journalists remarked upon the nationalism displayed by Chinese people online (e.g. Branigan 2008; Drew 2008;

D.K. Herold (✉)
Hong Kong Polytechnic University, Hong Kong, China
e-mail: ssherold@inet.polyu.edu.hk

P.-l. Law (ed.), *New Connectivities in China: Virtual, Actual and Local Interactions*,
DOI 10.1007/978-94-007-3910-9_3,

Forney 2008; Kennedy 2008). Within these stories, 'Cybernationalism' is often presented as an irrational outburst of China's 'fenqing', China's 'angry youth' whose oversensitive and misguided feelings for China are more the expression of Chinese insecurities rather than justifiable outrage over genuine offenses against the Chinese people. Consequently, their outbursts can be belittled and do not have to be accepted as a legitimate expression of widely held sentiments of the Chinese people (e.g. Osnos 2008).

This chapter wants to argue, though, that this is a misperception and misrepresentation of the growing nationalistic feelings in Chinese cyberspace and by extension among Chinese people in general. While there are many 'angry youth', who display extreme reactions to every perceived slight of their beloved China, there are increasing numbers of 'reflective nationalists' as well, who post well-argued entries on blogs and bulletin boards (BBS), calling for China to become stronger and to go its own way nationally and internationally.

The emergence of strong feelings of nationalism during the past 10 years should more justifiably be attributed to China's by now proverbially fast economic development and its attendant improvements to the lives of many young Chinese. The Chinese Communist Party (CCP) has been highly successful in equating the economic success with its own continuing rule of China, which has led to widespread support for the government among young Chinese. This support has disappointed many American and European politicians and journalists who expected China's economic development to culminate 'naturally' in the collapse of the Communist regime and the creation of a 'free', 'democratic' China with a multi-party state to govern it (see e.g. Miles 2007; *Spiegel Online* 2007; Tisdall 2007; Watts 2007).

Instead of fighting against the CCP for democratic reforms, China's Internet users have repeatedly shown themselves to be very patriotic and supportive of China and its future development, and this is likely to continue as long as the Communist party manages to improve the perceived state of the Chinese economy and China's status in the world (see MacKinnon 2008a). China's 'online citizens', or 'netizens' ('[inter-]net' + 'citizens'), approve of the strengthening of China, which means they approve of the Chinese Communist Party as long as it is improving the country both internally as well as on the international stage.

Before discussing some of the introspective and serious nationalists in Chinese cyberspace in this chapter, in the next part, I want to give an overview over some of the more noticeable postings by China's angry youths. This can serve as a counterpoint and background for the presentation of the serious arguments for Chinese Cybernationalism, as the latter should not be understood within a Euro-American frame of reference, but seen as reaction to the noisy nationalism of the fenqing.

2 Angry Young Nationalists

The term 'fenqing' was originally used for the Red Guards of the time of the Cultural Revolution and first appeared in the 1970s. It is an abbreviation for the term 'fennu qingnian' (angry youth) and is today mostly used as a derogatory term online and by journalists (see e.g. the references in the introduction) to refer to the stereotypical Chinese Internet user: male, young, nationalistic, full of anger and frustration (see Osnos 2008).

These young, angry, and frustrated Chinese react with a display of national pride and wounded sensibilities whenever they perceive China to have been insulted. This happens, in particular, on websites where Chinese netizens interact with Internet users from other countries. Examples of such pride can be found in many of the posts on the websites of Western newspapers, e.g. the Guardian Online site, or the sites for the German newspapers Die Welt or Der Spiegel. Any hint of criticism of China attracts attacks by Chinese posters, who appear to trawl the Internet looking for offending materials. These attacks are not restricted to news sites, though, as the group providing the free video software Videolan Client (VLC) found out. On their forums (*Videolan* 2006), a user from mainland China complained in 2006 that one of the Asian download mirrors for the free software was located in Taipei, Taiwan, and that this gave Taiwan the status of an independent country. The Chinese poster proceeded to educate the American programmers that Taiwan was only a province of China and that they should correct the wrong impression their website created, and the protests only ended when the frustrated American programmers locked down the forum thread.

If the offending non-Chinese happens to be in China while committing his 'crimes against China', the attacks can get a lot worse, as e.g. an English teacher with the online name of Chinabounder found out in 2006. Chinabounder had established a blog on the Blogspot blogging service to write about the author's sexual encounters with numerous Shanghainese women in great detail. He also used comments made by his female partners to criticise China, Chinese culture, Chinese politics, Chinese men, etc. on a wide range of issues and defended his views against all who attempted to argue against him in comments left on his blog (Chinabounder, n.d. – The entries referred to here date to the time before September 2006).

As Blogspot had been blocked in China at the time, he remained unnoticed by Chinese netizens for many months, although he had fast become a household name especially among male expatriates living in China who started chatting online and off-line about him. Once Blogspot was unblocked in China, though, this changed dramatically. Within a week, increasing numbers of Chinese netizens visited the site and started venting their anger about the blog in Chinese cyberspace.

In late August 2006, Prof. Zhang Jiehai, a professor of psychology at the Department of Sociology in the Shanghai Academy of Social Sciences, called for all Chinese to join him in a hunt of this immoral and racist foreigner and asked for help in identifying him so the authorities could expel him from China (Soong 2006; Chen 2006). The open call for help from Chinese netizens in hunting down a foreigner who was perceived as insulting all Chinese gained a lot of support on the Internet and resulted in widespread Western media attention as well (see e.g. Coonan 2006; *Voice of America* 2006; Watts 2006). The online hunt slowed down very fast, though, once Chinabounder appeared to have left China and stopped blogging on his site.

In 2008, during the Tibetan unrest in March 2008, many Chinese felt that the reports about Tibet in the Western media contained mistakes and misrepresentations and portrayed the situation as the brutal suppression by the Communist Chinese state of a non-aggressive, peaceful movement led by Tibetan monks.

> Rao Jin, 24, the founder of a small technology company in Beijing, said he was so angry about what he sees as foreign journalists' prejudice against China that last week he created a Web site, http://www.anti-cnn.com, to document what he calls mistakes and bias in Western media. He said more than 1,000 people have e-mailed, volunteering to spot errors. (Drew 2008)

Anti-CNN was started in late March 2008 as a single web page with pictures of Western news programmes, reports, and web pages and attempted to document the mistakes they contained (the website has changed since then, but their initial reports have been discussed, e.g. by China Daily 2008; MacKinnon 2008b, March 26; and several of the entries in Soong 2008a, March). While Anti-CNN has changed since then (see MacKinnon 2009), the original web page led with a statement that expressed Rao Jin's disgust with Western reporting standards:

> See the true despicable and shameless face of western media. For a long time now, certain western media best represented by CNN and BBC, in the name of press freedom have been unscrupulously slandering and defaming developing nations. In order to achieve their unspoken goal they mislead and they ensnare, switching black for white, confusing right and wrong, fabricating…willing to go to any length. In their reports on the riots in Tibet Western media's performance once again shows to the world their repulsive true face. (Kennedy 2008 – also the following quotes)

This was followed by a call to arms to all Chinese netizens who were asked to join in and to track down other falsehoods in Western media reports '*not limited by language, content (text or photos) or country*', as '*the more evidence of their crimes we collect, the more space we'll have fought and won for ourselves*'. The enthusiastic response by young Chinese netizens to this call to arms demonstrated within a few days how angry they were with 'the West' and particularly with the 'Western' media. The introductory statement closed with a wider 'declaration of war':

> This is a struggle of resistance against western hegemonic discourse. We need to fully recognize that this will be a long-term, difficult and complex battle. But regardless of the outcome, we all firmly believe: Western nations' days of using several of their crap media in an absurd attempt to fool people with their rotten words will soon be over for good! (Kennedy 2008 – he also uses the phrase 'declare war')

As Western journalists living in China discovered, many young Chinese took this appeal to stand united and to protect China against all attacks very seriously (see e.g. Associated Press 2008).

In a similar vein, on 15 April 2008, '*a twenty-eight-year-old graduate student in Shanghai named Tang Jie*' (Osnos 2008) uploaded a short video clip to a video-sharing website entitled '2008 China Stand Up!' The video was later cross posted to YouTube, where it was watched over 50,000 times and attracted over 1,500 mostly positive comments (see Clover19862003 2008, for the video and the comments of many angry Chinese netizens).

> It was a homespun documentary, and it opened with a Technicolor portrait of Chairman Mao, sunbeams radiating from his head. Out of silence came an orchestral piece, thundering

> with drums, as a black screen flashed, in both Chinese and English, one of Mao's mantras: 'Imperialism will never abandon its intention to destroy us'. Then a cut to present-day photographs and news footage, and a fevered sprint through conspiracies and betrayals. [...] A cut, then, to another front: rioters looting stores and brawling in Lhasa, the Tibetan capital. The music crescendos as words flash across the scenes: 'So-called peaceful protest!' A montage of foreign press clippings critical of China – nothing but 'rumors, all speaking with one distorted voice'. [...] 'Obviously, there is a scheme behind the scenes to encircle China. [...] One final act of treachery: in Paris, protesters attempt to wrest the Olympic torch from its official carrier, forcing guards to fend them off – a 'long march' for a new era. The film ends with the image of a Chinese flag, aglow in the sunlight, and a solemn promise: 'We will stand up and hold together always as one family in harmony!' (Osnos 2008)

As a reaction to the Tibetan unrest and the disturbances during the Olympic Torch Relay, many Chinese netizens protested against the unfair treatment they thought China was receiving from 'Western' countries. These feelings grew on the Internet and later affected events off-line, as e.g. Cheng (2009) showed in his article on the 2008 Carrefour boycott in China.

Throughout 2008, there were a number of further sudden, nationalistic outbursts in Chinese cyberspace that at times affected off-line life as well (e.g. Leroux 2008), and even in 2009, there were a few flare-ups (e.g. Chen 2009; Fauna 2009). The pattern is very similar in each case, as it usually starts with a (group of) Chinese netizen(s) encountering something they see as an outrage and an insult to China which they then feel compelled to blog or post about. Their post is seen by other young Chinese, who in turn are angered and cross post the original posting to other blogs and BBS, at which point even minor events become topics hotly debated by hundreds of thousands of Chinese netizens. Through the sheer volume of discussions, others are alerted to the topic, which gains in importance with each new post online, until people's attention turns to the next piece of outrage. Compared to the numbers involved, though, the actual effect of most of these flare-ups and their impact on life in China is insignificant. Although individual lives are sometimes impacted negatively, the amount of emotional energy invested into these online debates and the size of the netizen population involved in the discussions reduce these outbursts to the level of venting. Judged on their effectiveness, they appear more as 'steam valves' rather than as genuine expressions of deeply held beliefs.

3 'Reasonable Nationalism'

Compared to the hot air expended by the 'angry youths' of China's cyberspace, the number of posts by people posting 'reasonable' nationalistic arguments online is negligible. Their background, their standing, and the similarity of their argument, though, suggest that they present a far more problematic development of Chinese public opinion. They are not just young people venting their frustrations online or whiling away time by taking part in the latest round of 'foreigner bashing'. Instead, these posters took the time to sit down and reason through China's position in the world and came to very nationalistic conclusions based on their reasoning.

One of them, with the online name 'Hecaitou' is one of the most influential bloggers in Chinese cyberspace (see Xiao 2008b, December 7). During the Tibetan unrest in 2008, he posted an entry on his blog containing several images of protests against China in European cities and used this to attack Western '*prejudices against the Chinese people*' (see Soong 2008a, March, also for the following quotes). He argued that the protesters' '*efforts not only provide no material help, but they stir up the nationalistic fervor of the Chinese people*'. He accused Westerners of believing that '*the Chinese have been brainwashed*' and that '*all Chinese are ignorant, undeveloped and close-minded*' and turned the accusation around by pointing out '*many Chinese people know as much as they do and in fact visited a lot more websites than they have*'. He explained that many young Chinese access a variety of European and American news websites, while only very few non-Chinese have the language skills necessary to access Chinese sources. In this situation, he concluded it was pointless for Chinese people to attempt to interact with Europeans or Americans:

> In the absence of respect and equality, what is the point of dealing with the Westerners? Presently, the Westerners must be wondering about the reaction of the Chinese people to the current events. Once again, they treat the unexpected outcome as the result of successful brainwashing or overflowing nationalism. But they would never reflect on the implications of their actions on the Chinese people.

Hecaitou felt that Westerners treated him and other Chinese netizens with open contempt and refused to accept their opinions. Moreover, this unacceptable attitude of Europeans and Americans towards China and the Chinese people was only going to exacerbate the situation.

> Thanks to their concerns, the Chinese people have rallied at an unprecedented speed underneath the national flag. They have voluntarily given up many rights and freedoms, in order to avoid more injuries and insults from the outside. These Westerners are not helping their friends. They are only helping to create an enemy as well as an Asiatic orphan.

While Hecaitou's comments could be disregarded as being too sensationalist, and his blog as not being representative of 'the Chinese people', it has to be remembered that his blog entry was copied to many of China's major bulletin boards and discussed across Chinese cyberspace – and most Chinese netizens agreed with him.

One netizen who agreed with Hecaitou, and whose post was widely circulated in support of Hecaitou's comments, was a Chinese living in Germany who went by the name of 'Schweinsteiger'. On 1 April, Schweinsteiger posted a long entry on the ChinaRen BBS forum (see Soong 2008b, April 1 for a translation and the quotes below) in which he described the interactions and conversations he had with several colleagues in an office in Germany during the Tibetan unrest.

According to the post, his colleagues were eager to prove to him that the Chinese government was persecuting and suppressing the Tibetan people in Tibet. Using reports in the German media, they attacked China and the Chinese government, refusing to believe anything reported within China because '*your media are lying and what you see is not real! […] Your television channels are telling lies*'.

After recounting numerous interactions with his colleagues, 'Schweinsteiger' summarises what he believes are the main problems for China's image in Germany and, therefore, for any Chinese wanting to live or study in Germany:

> [...] 3. Chinese students in Germany are unpopular. My colleagues indirectly reveal those feelings. They even tell me directly: 'We are really worried about what happens if one day you learn what we know'.
>
> 4. It is a mistake for China to even exist. The faults of China can be stacked from the ground to the heavens. Furthermore, under the leadership of this demon government, things are getting worse and worse.
>
> 5. Bloodshed and massacres occur everywhere in China. When a Chinese citizen says the wrong thing, he will be arrested immediately and subjected to extreme torture in jail.

He concludes his post with several sentences expressing his frustration with the treatment he received from his colleagues. They are similar in tone to Hecaitou's statements, but take the idea a step further, arguing that China should no longer listen to other countries and should instead focus on growing stronger, so that one day it might get the respect it deserves.

> Anything good about China must have been fabricated by the Chinese government; visual images favorable to China were staged by the Chinese government; any photo favorable to China was the result of PhotoShop work. [...] China is hopeless with no redeemable value. All the opposing voices against China are right, and they will support those voices. [...] I feel that it is a long and endless struggle with them. This struggle cannot be resolved through any debate or discussion of facts. **This can only be done through the construction of the motherland. When the motherland is strong, even stronger, they will shut their mouths!** Each one of us Chinese overseas students is working hard and enduring the suffering. Several decades into the future, will China collapse like the Germans hope? Or will China be so strong that they will collapse? (My emphasis)

Half a year later, on September 2, 2008, a Chinese calling 'Mr. Li' submitted a letter of complaint to BBC China, which was later reposted widely across Chinese cyberspace (Soong 2008d, September 14 – also the following quotes). Mr. Li called himself pro-Western and stated that he used to support the idea of democracy in China. His attitude began to change, though, during a stay in eastern Europe during the 1990s as '*during the four or five years when I could only see the western media, all the reports that I saw about China were negative and critical*'. These reports were at odds with the fast developing China he encountered during his visits home. During the following years, he began to believe that Western media were deliberately misrepresenting China in their reports and he became especially frustrated while watching the misrepresentation of China during the months leading up to the Olympic Games in 2008.

> The western media may not realize that they are losing China! They are losing the admiration and trust of the young generation of China, pushing them towards nationalism. All this occurs because the western media do not really understand China and they have no intention of really understanding China either.

In Mr. Li's opinion, Western media were projecting East European labels onto China, believing '*that the Chinese government is a dictatorial and totalitarian government, which must necessarily be unpopular among the people*'. Answering this accusation, he claimed to speak for the '*middle-class and intelligentsia that came into being*' since the reforms in China started in the late 1970s. He argued that the Chinese '*want China to go even further with* [economic] *reforms*' and affirmed that the middle class '*basically support the Chinese government*'. He argued that '*democracy and the rule of law are the ultimate goals of modernization for China*', but that at present, it was more important for China to '*maintain a strong and powerful central government*'.

Mr. Li criticised Europe and America, saying that they '*seem to want democracy for the sake of democracy and they don't care what happens to China after democracy and freedom come*', and he accused them of using '*democracy and freedom as pretexts to divide and weaken China*' so that China would not rise up to challenge the current status quo. This follows the same train of thought as the arguments presented by Hecaitou and Schweinsteiger, but adds the notion that other countries are deliberately attempting to sabotage China and the continued development of the PRC.

About the same time that Mr. Li's essay was published, a netizen by the name of '300 Spartan Heroes' posted an essay with the title 'How the Western Media Lost the Young Generation in China' on the popular Tianya bulletin board (see Soong 2008c, September 11 for a translation and the quotes below). The essay was similar to the posts discussed above, but in an interesting twist: it seemed to receive the blessing of the Chinese government through the so-called 50 cent gang (Wu-Mao-Dang), pro-government bloggers who are suspected of commenting on bulletin boards and blogs to further the agenda of the central government (see e.g. Xiao 2008a, October 28). Most of the comments left for the essay were extremely supportive, while also praising the quality of Chinese TV productions. Other commentators pointed to the ludicrousness of such statements, and the debate in the comments soon turned against the '50 cent gang' (some of the comments can be found in translation in Pan 2008).

Nevertheless, the original essay made a number of important points in the development of nationalistic ideas in Chinese cyberspace and should not be ignored. The essay point of departure was the assertion that '*Western nations*' are intent on '*promoting their values*' through the '*hard methods*' of waging wars (e.g. Afghanistan and Iraq) and the '*soft methods*' of the '*Western media using their international speech rights to say awful things about countries which do not have Western-style democracy*'. This information war worked in the past, but has recently started to fail as increasing numbers of people have begun travelling to and from China.

> If there are no western tourists coming to China and no Chinese studying overseas, the western media could say whatever they want [...]. If you cannot see for yourself, you have to trust them. But times are different, as more and more western visitors come to China and more and more Chinese tourists travel overseas. [...]The western tourists are perplexed because China is completely unlike what their own media are reporting. The overseas Chinese students are perplexed because very few western media reports have anything good to say about China. [...] Those who have seen the real China realized that they had been deceived by the western media.

Making a similar point as Mr. Li, '300 Spartan Heroes' continued this line of thinking by asserting that this discrepancy between 'Western' media reports and the experienced reality was particularly felt by China's new middle class and the intelligentsia who have profited the most from the continuing economic development.

> Over the past three decades, the Chinese government has led the country to an astonishing economic growth, and many citizens have benefited from it. The Chinese who travel overseas during this period are the rapidly rising middle class and the intelligentsia. When they see the good things in China being badmouthed in the western media, what else is this but hypocrisy?

The essay concluded with the idea that both the middle class as well as the intelligentsia would like to live in a democratic China, but that a democratic system for China was not as important as the well-being of the Chinese people and the development of the Chinese economy (see similar findings by Wilson 2009, in his study of reactions to film censorship in China). Using actions of the USA in Iraq and of France and Great Britain in Africa as illustrations for his point, '300 Spartan Heroes' argued that

> ultimately, the Chinese people want to achieve prosperity and national power through democratization. But the Western media seem to only want democracy for the sake of democracy and they don't care what happens to China afterwards. The Chinese form of democracy guarantees first and foremost the right to survive and develop. But the Western media wants to promote its own form of democracy according to its own ideas. They don't care what happens to a country afterwards.

The notion that there is a Chinese '*form of democracy*' that requires a strengthening to the country before a relaxation of control can happen has since become very influential in China, and in many current debates, the idea that China needs a political system based on 'Chinese' values and on the specific necessities of Chinese culture is used to oppose calls for democratisation, liberalisation, or the creation of a multi-party system. Just as the Chinese Communist Party has successfully run the country using an economic system that combines Capitalist and Communist elements to create a 'Socialist Market Economy with Chinese Characteristics', so it is now proposed that China should develop a uniquely Chinese form of democracy that takes into account China's 'special needs'.

Based on the postings of people like Hecaitou, Schweinsteiger, Mr. Li, and '300 Spartan Heroes', the construct of a 'Chinese democracy' would include notions of a free market economy and the control (and safety in the streets) of a dictatorship, reminiscent of the political system of Singapore. China would continue to grow stronger and would be accepted and respected as a world power by other countries. Attacks (verbal and others) on China by non-Chinese would cease, as people in Europe and America would recognise China's strength, etc. As Wu (2007) pointed out, Chinese Cybernationalists believe in a curious mixture of freedom online, a (free) market economy, strict political controls, the revival of (Han) Chinese culture and traditions, and 'proper' Chinese morals (see also Jacques 2009; Chu and Cheng 2011).

> It's important that the outside world understand that China's patriotic youth [...] feel that they are acting on their own belief systems and get angry when characterized as brainwashed puppets. (MacKinnon 2009)

Additionally, it should be noted that this notion of 'Chinese democracy' revolves around the idea that other countries will accept China as equal. The 'reflective nationalists' in China's cyberspace are arguing **for** democracy, but only if this democracy is homegrown and free from European or American patronising. They are very (over) sensitive to the way in which American and European politicians, academics, media, etc. offer to 'help' China to become more open, more democratic, and more respectful of human rights:

> The Westerner stoops down condescendingly to stretch out a helping hand to the wretched little yellow men so as to educate and instruct them. They are totally oblivious to the possibility that they are dealing with live human beings who are thoughtful and sentient. (Hecaitou as translated in Soong 2008a, March)

A discussion of Western attitudes towards China is beyond the scope of this chapter and would require detailed analyses of many forms of communications towards and about China. Whether the comments of Chinese netizens about Western media and 'the West' in general are accurate or not, though, their postings demonstrate that many Chinese online feel continuously and unjustly threatened by the non-Chinese world, unfairly presented and accused, and have come to resent what they see as the 'Western attitude' towards China and its development.

> China faces a hegemony of media discourse that perpetuates its already negative image by focusing on China's sordid human rights records, its difficulties managing the quality control of manufactured goods and in securing intellectual property rights, and its pollution and environmental degradation. (Hagan 2008: 77f)

They are not traditionalists fighting a globalising development bent on slowly destroying their cultural heritage (Farrer 2007; Yuan 2007), but rather proud citizens of the world's fastest developing country exasperated with the constant criticism of the self-declared arbiters of the world in the USA and Europe. They feel contempt for the hypocrisy of the USA and the political weakness of Europe and prefer to rely on the proven economic (and political) strategies of the Chinese Communist Party rather than to stage a revolution and to descend into the chaos that was Russia of the 1990s.

4 China Is Unhappy – And the World Should Take Notice

On 12 March 2009, five Chinese academics (Song et al. 2009) released a book with the title *Unhappy China: The Great Time, Grand Vision and Our Challenges* containing a collection of essays calling upon Chinese people to have more confidence in their own country and upon China to start conducting its foreign policies from the position of strength the rapid development of the past three decades had earned it.

While the book has been widely criticised by academics, netizens, and other Chinese people (see e.g. Li 2009), it echoes similar sentiments to what netizens have posted online during the past few years. The book also demonstrates that many Chinese are dissatisfied with China's position in the world and wish to see it change, something pointed out by Xiao Gongqin, professor of history at Shanghai Normal University, in an interview with Xinmin Weekly:

> The question of modern China's relationship with other peoples of the world [...] tugs at people's hearts because 30 years of reform have made the Chinese people stronger, and so after a century of shame Chinese people face the question of how to re-define ourselves. [...] I believe that for some time to come this nationalist wave as epitomized by Unhappy China will continue to exist, and foreigners will have to learn to come to terms with this [...] form of Chinese nationalism. (Translated in Bandurski 2009)

Across Chinese cyberspace, the book is hotly debated, and the book's text can be downloaded from thousands of sites. Even if the book may only have been the cynical attempt to instrumentalise the anger of many young Chinese to make a fast profit (see the book review by Jing Kaixuan translated in Bandurski 2009, April 2), its impact is undeniable. The book seems to have judged the mood of the majority of the Chinese people accurately with its title, even if the arguments presented and the enemies chosen are often seen as weak and/or misguided.

> When I first heard about the book Unhappy China, I thought it was probably about how laid-off workers were unhappy, or about how peasants who had lost their land were unhappy. Maybe it was about how college graduates searching for work were unhappy, about how stock market investors were unhappy, or about how victims of the poisonous milk powder scandal were unhappy. (Jing Kaixuan translated in Bandurski 2009)

All across China, both online and off-line, academics critical of the lack of social reforms, nationalistic youths, disgruntled workers and entrepreneurs angry about official corruption, environmental activists, etc. are complaining about and trying to find solutions for China's problems (see e.g. Hartono 2009, April 20), but despite all the problems they see and raise, they still feel connected to China and are proud of being Chinese, and they resent foreign labelling (see e.g. Cha 2009) or unsolicited advice (see e.g. Ash 2008).

Whether (off-line) China will follow the arguments of the netizens of Chinese cyberspace and become more nationalistic in its relations with other countries will depend neither on a resurgence of traditional Chinese culture nor on the machinations of the CCP, but to a large extent on the treatment China is accorded by the rest of the world. Real or imagined, China's netizens and citizens resent foreign slights against their country and want the world to respect that.

References

Ash, T. G. (2008, December 4). China's economic success may soon bring trouble. It would be ours too. *The Guardian*. From http://www.guardian.co.uk/commentisfree/2008/dec/04/garton-ash-china-economy-comment. Accessed 29 Apr 2011.

Associated Press. (2008, April 14). Journalists group appeals to China to allow unrestricted coverage around Olympics. *ESPN*. From http://sports.espn.go.com/espn/wire?section=oly&id=3347201. Accessed 29 Apr 2011.

Bandurski, D. (2009, April 2). Unhappy China, and why it is cause for unhappiness. *China Media Project Blog*. From http://cmp.hku.hk/2009/04/02/1544. Accessed 29 Apr 2011.

Branigan, T. (2008, March 23). Western stance infuriates young Chinese fed on government propaganda. *The Observer*. From http://www.guardian.co.uk/world/2008/mar/23/tibet.china. Accessed 29 Apr 2011.

Cha, A. E. (2009, April 19). For China's New Left, old values. *The Washington Post*. From http://www.washingtonpost.com/wp-dyn/content/article/2009/04/18/AR2009041801939_pf.html. Accessed 29 Apr 2011.

Chen, P. (2006, August 31). Shanghai + sex + blog + controversy = book deal? *Shanghaiist*. From http://shanghaiist.com/2006/08/31/we_gonna_go_to.php. Accessed 29 Apr 2011.

Chen, B. (2009, January 2). China: "Just the aftershocks from the earthquake would destroy France!" kids yell. *Global Voices Online*. From http://globalvoicesonline.org/2009/01/02/china-just-the-aftershocks-from-the-earthquake-would-destroy-francekids-yell/. Accessed 29 Apr 2011.

Cheng, C. (2009). New Media and event: A case study on the power of the Internet. *Knowledge, Technology & Policy, 22*(2), 145–153.

China Daily. (2008, April 3) CNN: What's wrong with you? *People's Daily Online*. From http://english.people.com.cn/90001/90776/90883/6386008.html. Accessed 29 Apr 2011.

Chinabounder. (n.d.). *Sex and Shanghai*. From http://chinabounder.blogspot.com/. Accessed 29 Apr 2011.

Chu, W.-C. R., & Cheng, C.-T. (2011). Cultural convulsions: Examining the Chineseness of cyber China. In D. K. Herold & P. Marolt (Eds.), *Online Society in China: Creating, celebrating, and instrumentalising the online carnival* (pp. 23–39). London/New York: Routledge.

Clover19862003. (2008, April 17) 2008! China Stand Up! *YouTube*. From http://www.youtube.com/watch?v=MSTYhYkASsA. Accessed 29 Apr 2011.

Coonan, C. (2006, September 4). China's internet vigilantes target British ex-pat cad. *The Independent*. From http://www.independent.co.uk/news/world/asia/chinas-internet-vigilantes-target-british-expat-cad-414557.html. Accessed 29 Apr 2011.

Drew, J. (2008, March 24). Protests may only harden Chinese line. *Washington Post*. From http://www.washingtonpost.com/wp-dyn/content/article/2008/03/23/AR2008032301595_pf.html. Accessed 29 Apr 2011.

Farrer, J. (2007). Asian youth culture in a globalizing world. *Global Asia, 2*(1), 102–110.

Fauna. (2009, April 15). "Zhejiang University girl" exposed by Human Flesh Search. *Chinasmack*. From http://www.chinasmack.com/2009/stories/zhejiang-university-girl-exposed-by-human-flesh-search.html. Accessed 29 Apr 2011.

Forney, M. (2008, April 13). China's loyal youth. *The New York Times*. From http://www.nytimes.com/2008/04/13/opinion/13forney.html. Accessed 29 Apr 2011.

Grossman, L. (2006, February 5) Google under the gun. *Time*. From http://www.time.com/time/magazine/article/0,9171,1156598-1,00.html. Accessed 29 Apr 2011.

Hagan, C. P. (2008). One world, one dream? – Image building and brand China in the 2008 Beijing Olympics. MA thesis, University of Oregon, Eugene.

Hartono, P. (2009, April 20). Liu Jianqiang, Zhang Ping, Wang Lixiong: On the impact of New Media. *China Digital Times*. From http://chinadigitaltimes.net/2009/04/liu-jianqiang-zhang-ping-wang-lixiong-on-the-impact-of-new-media/. Accessed 29 Apr 2011.

Jacques, M. (2009). *When China rules the world: The end of the western world and the birth of a new global order*. London: Penguin Press.

Kennedy, J. (2008, March 24). China: Bloggers declare war on Western media's Tibet coverage. *Global Voices Online*. From http://globalvoicesonline.org/2008/03/24/china-bloggers-declare-war-on-western-medias-tibet-coverage/. Accessed 29 Apr 2011.

Leroux, M. (2008, April 22). Grace Wang targeted by Chinese internet users. *The Times*. From http://www.timesonline.co.uk/tol/life_and_style/education/student/news/article3797519.ece

Li, L. (2009, April 16). Happy or unhappy is the question. *Beijing Review*. From http://www.bjreview.com/nation/txt/2009-04/13/content_190569.htm. Accessed 29 Apr 2011.

MacKinnon, R. (2008a). Flatter world and thicker walls? Blogs, censorship and civic discourse in China. *Public Choice, 134*, 31–46.

MacKinnon, R. (2008b, March 26). Anti-CNN and the Tibet information war. *RConversation*. From http://rconversation.blogs.com/rconversation/2008/03/anti-cnn-the-me.html. Accessed 29 Apr 2011.

MacKinnon, R. (2009, April 14). My chat with Anti-CNN.com. *RConversation*. From http://rconversation.blogs.com/rconversation/2009/04/on-monday-afternoon-i-did-an-online-chat-with-these-patriotic-young-people-who-run-anti-cnn-a-website-launched-in-the-wake.html. Accessed 29 Apr 2011.

Miles, J. (2007, December). The challenge to Beijingoism. *The Economist*. From http://www.economist.com/theworldin/china/displayStory.cfm?story_id=10125628&d=2008. Accessed 29 Apr 2011.

Osnos, E. (2008, July 28). Letter from China: Angry Youth. *The New Yorker Online*. From http://www.newyorker.com/reporting/2008/07/28/080728fa_fact_osnos/

Pan, K. (2008, September 14). Chinese Youth on Western Media: A diversity of opinion. *CNReviews.com*. From http://cnreviews.com/china_blogosphere/chinese_youth_on_western_media_a_diversity_of_opinion_20080914.html. Accessed 29 Apr 2011.

Song, X., Wang, X., Huang, J., Song, Q., & Liu, Y. (2009). *Zhongguo bu gaoxing*. Beijing: Kong Hong Book Company.

Soong, R. (2006, August 28). The immoral foreign blogger. *EastSouthWestNorth*. From http://www.zonaeuropa.com/20060828_1.htm. Accessed 29 Apr 2011.

Soong, R. (2008a, March). Daily brief comments March 21–31, 2008. *EastSouthWestNorth*. From http://www.zonaeuropa.com/200803c.brief.htm. Accessed 29 Apr 2011.

Soong, R. (2008b, April 1). Three days with a German. *EastSouthWestNorth*. From http://www.zonaeuropa.com/20080403_1.htm. Accessed 29 Apr 2011.

Soong, R. (2008c, September 11). Daily brief comments – 03: How the Western Media lost the young generation in China. *EastSouthWestNorth*. From http://www.zonaeuropa.com/200809b.brief.htm. Accessed 29 Apr 2011.

Soong, R. (2008d, September 14). Mr. Li writes to BBC. *EastSouthWestNorth*. From http://www.zonaeuropa.com/20080914_1.htm. Accessed 29 Apr 2011.

Spiegel Online. (2007, August 8). Olympia-Party in Peking: Countdown für die Menschenrechte. *Der Spiegel*. From http://www.spiegel.de/sport/sonst/0,1518,498907,00.html. Accessed 29 Apr 2011.

Tisdall, S. (2007, October 15). Beijing bluster. *The Guardian*. From http://www.guardian.co.uk/commentisfree/2007/oct/15/beijingbluster. Accessed 29 Apr 2011.

Videolan. (2006, August). A serious problem on the page of downloading VLC!!!!!!!! From http://forum.videolan.org/viewtopic.php?f=10&t=25748. Accessed 29 Apr 2011.

Voice of America. (2006, September 4). Internet users turn vigilante in China. From http://www.voanews.com/english/news/a-13-2006-09-04-voa12.html. Accessed 29 Apr 2011.

Watts, J. (2006, August 30). Webhunt for 'British bounder'. *The Guardian*. From http://www.guardian.co.uk/technology/2006/aug/30/news.newmedia1. Accessed 29 Apr 2011.

Watts, J. (2007, October 14). All change as Chinese face game of their life. *The Observer*. From http://www.guardian.co.uk/world/2007/oct/14/china.jonathanwatts. Accessed 29 Apr 2011.

Wilson, M. (2009). Transgressive conformity: DVD and downloading in China. *International Journal of Zizek Studies [Online], 3*(1). From http://www.zizekstudies.org/index.php/ijzs/article/view/177/273. Accessed 29 Apr 2011.

Wu, X. (2007). *Chinese cyber nationalism: Evolution, characteristics, and implications*. Lanham: Lexington Books.

Xiao, Q. (2008a, October 28). Authorities' attempts to bring online public opinion under control. *China Digital Times*. From http://chinadigitaltimes.net/2008/10/cyber-police-try-to-avoid-mass-incident-with-a-preemptive-post/. Accessed 29 Apr 2011.

Xiao, Q. (2008b, December 7). China News: Twenty most influential figures in China's cyber-space. *China Digital Times*. From http://chinadigitaltimes.net/2008/12/twenty-most-influential-figures-in-chinas-cyberspace/. Accessed 29 Apr 2011.

Yuan, W. (2007). Nationalism in a transforming China. *Global Asia, 2*(1), 21–27.

Part II
Studies on Mobile Phone Uses in Beijing

Chapter 4
Beijing Calling… Mobile Communication in Contemporary China

Leopoldina Fortunati, Anna Maria Manganelli, Pui-lam Law, and Shanhua Yang

1 Modernization and the Social Effects of New Media in China

Since the late 1970s, economic and social reforms in China have set in motion radical transformations. The "open door" policy to accept direct investments from abroad (initially, from the 50 million Chinese who constitute the "dragon diaspora" throughout the world) and industrial development have led to a resurgence of the Chinese economy (Weber 2003). Since it offers a low-cost labor force, China has become attractive, first, to multinationals and, then, to mid-size firms and even small enterprises in a wide range of sectors. Today, China is considered the "factory" of the globalized world. The modernization of China, however, is not confined to the provision of production at a low added value: it also entails technological appropriation and innovation. In particular, China has become an important world producer of information technologies—the hardware, software, and equipment for telecommunications (Weber 2003). Given the massive government investments that are flowing into this sector, there is no doubt that China will soon become one of the world's leading ICT nations. In the meantime, it represents the biggest world market for mobile phones and will soon dominate the Internet market. In China, the total number of mobile and fixed-line subscribers exceeds 750 million, and the total

L. Fortunati
Faculty of Educational Sciences, University of Udine, Udine, Italy

A.M. Manganelli
Department of General Psychology, University of Padova, Padova, Italy

P.-l. Law (✉)
Department of Applied Social Sciences, The Hong Kong Polytechnic University, Hong Kong, China
e-mail: sspllaw@inet.polyu.edu.hk

S. Yang
Department of Sociology, Peking University, Beijing, China

P.-l. Law (ed.), *New Connectivities in China: Virtual, Actual and Local Interactions*,
DOI 10.1007/978-94-007-3910-9_4, © Springer Science+Business Media B.V. 2012

number of Internet users is more than 137 million. According to Jiang Yaoping, vice-minister of Information Industry, the ICT sector now represents a leading industry in the Chinese economy. In the last few years, the government has implemented the "Village Connected Project," to improve the telephone access rate in rural and remote areas of China. At the end of 2005, about 97.1% of the villages in China were connected by at least one main line.

The industrial and commercial development of China was accompanied and made possible by rapid changes in the social structure of the nation, including the decrease in population growth, the growth of urbanization following the relaxation of controls on internal migration at the end of the 1990s, the harrowing division of millions of families from the inner regions, the increased demand for goods, and the explosion of the real estate market following the rapid rise of building construction. These changes have been reinforced by governmental policies such as the "break of the iron bowl" policy (which marks the end of the guaranteed job) and strategies to stimulate mass ownership (beginning with homes). These events have set in motion powerful processes of social stratification (Tomba 2003). In China today, there has been a domestic migration of more than 225 million peasants from poor regions in the interior to the coastal areas, especially in the south. These migrants, who are employed by industries belonging to the new Chinese capitalist system, have fewer civil rights than the locals. At the same time, there has been a development of an urban middle class (employees, civil servants, technicians, professionals, political managers, executives, etc.), an upper class (entrepreneurs, trading professionals, etc.), and a very affluent class. In all classes, women's social position is generally inferior to that of men (Yuen et al. 2004).

What effects have social stratification and the adoption and spread of the mobile phone and the Internet had on the greater part of the Chinese population? To what extent are the new media endorsing modernization? What are the social consequences of the use of ICTs on the organization of families, the structure of social relationships, the relation between towns and the countryside, the public sphere, and the organization of everyday life in China? Such questions have attracted a considerable amount of attention: qualitative research studies have already explored some of the social consequences of the rapid diffusion and adoption of the mobile phone in contemporary China, particularly in the capital, Beijing, and coastal industrialized towns.

For example, according to Law and Peng (2006), the adoption of the mobile phone by migrant workers in Guangdong is critical to the understanding of the symbolic meaning of this device. The workers spend 3 or 4 months' wages to buy a mobile phone and a considerable part of their monthly income to use the device. Law and Peng (2006) also examine the social effects of domestic migration on the structure of family relationships. They depict the changing attitudes of recent generations of migrant workers toward their families of origin. These workers use their wages for personal needs, rather than improving the welfare of their families. Mobile phones allow migrant workers to loosen their family ties and develop a higher level of individualization. According to Law and Peng (2005), migrant workers develop a new interest in commodities and modern lifestyles in response to the difficult working conditions in factories. In prosperous Chinese cities, modern lifestyles are supported by industrialization and the globalized market.

New media are often considered "technologies of freedom" (de Sola Pool 1988) because they are supposed to enhance communication and expand the personal freedom of users. Law and Peng (2004a, b, 2005, 2006; see also Chu and Yang 2006), on mobile phone use by migrant workers, show how this device empowers users in various ways: it allows them to access job information and so improves their working conditions, to keep in contact with their family at home, to enlarge their social sphere through connections in the cyberspace, and to expand their concept of leisure.

Another qualitative research study supports the view that in China, modernization strengthens the process of individualization. Yang (2005) surveyed a group of Beijing inhabitants and found that while mobile communication may help to maintain parent/child ties and facilitate interactions between needs- or interests-based friends, it does not reinforce ties such as those between best friends or couples. Fortunati and Yang's (2008) study of everyday mobile phone use demonstrates that users in China put this new technology to more advanced use than their counterparts in Europe.

Quantitative research on the use of mobile phones, however, is completely lacking. Such research is necessary to determine the basic structure and dimension of the diffusion, adoption, and use of mobile phones in contemporary China. Of course, only quantitative research that is based on a representative sample of Chinese population would be able to fill this gap. However, such research is unavailable for a number of reasons such as the issue of the accuracy of the official demographic data, the high cost of face-to-face survey for a representative sample, and so on and so forth. In this study, we undertook quantitative research, but our sample group was a convenience sample of 487 inhabitants in Beijing. Although this sample can hardly be used as a basis for generalizations about the attitudes, behavior, and use patterns of mobile phone users in Mainland China, it might serve as an exploratory attempt to understand the general pattern of the mobile use in China. That is, it might be considered a kind of pretest of a future research.

The present study focuses on how, after a decade of mobile phone use, the inhabitants of Beijing evaluate the changes in the social and communicative sphere as a result of the introduction of this technology.[1] It addresses the following research questions: After a decade of mobile phone use, how do Chinese people perceive the importance of this device? To what extent do the adoption and use of mobile phones increase or decrease social connectivity in contemporary China? To what extent does the use of mobile phones in everyday life enhance or reduce the communications—do they make you feel closer to or more distant from other people? What are the variables that predict users' attitudes toward mobile phones in China?

[1] For spatial reasons, it was not possible to present in this chapter also the results related to behavior and practices of use of the mobile phone. This data have been illustrated and discussed in the following papers: Fortunati et al. (2008, 2009, 2010).

2 Aim and Method

The purpose of the study is broad and covers a number of aspects of mobile phone practices in China. In this chapter, however, we address only the research questions listed above. The data are the result of a face-to-face survey designed and conducted in March 2006 in Beijing. Demographic variables included gender, age, education, income, and place of origin (rural or urban). The participants ($n=487$) were Beijing inhabitants; 280 were males (57.5%) and 207 females (42.5%). The respondents' ages varied from 15 to 60 years, but the large majority of the sample (86%) was under 30. Almost half of the respondents were born in the countryside. Nearly a quarter had only a basic level of education (secondary school graduation), one-fifth had some further education (high school or some years of college), and more than half had at least one degree. With respect to income, 21.6% of the sample had no income; 30.6% made up to 1,000 *yuan* per month, 34.3% made from 1,001 to 4,000 *yuan*, and 13.1% over 4,000 *yuan*.[2]

3 Results

In response to questions about the perception of connectivity after the advent of the mobile phone, 81% of the respondents declared that the mobile phone increased their connectivity, while 18.6% thought that their number of their contacts remained stable. This very clear result leads us to conclude that, for these respondents, the immediate appraisal of the social effects of mobile phone use on their network of personal contacts is unambiguously positive. None of the demographic variables created a significant difference in these responses.

The general perception among our sample of an increase in their contact with others is reinforced by the fact that the majority of respondents (62.4%) believed that mobile phone use has enhanced the sphere of their social communication. For respondents born in rural areas, this percentage rose to 72.2%, and for those with low-paying jobs (under 1,000 *yuan*), it rose to 75.8%.[3] The respondents who perceived an enhancement of their social circle were also more likely to have a low level of education (77.5%).[4] Those who felt that their communication network remained unchanged or even decreased were predominantly women (44.7%, versus 32.2% of men).[5] Age was not a factor in these answers. For the majority of respondents, then, the mobile phone was perceived as a tool that enhances connectivity and widens communication and social networks. It is worth noting that this is particularly

---

[2] One euro is about 10 *yuan*.

[3] $\chi^2_{(1)}=19.33, p<0.0001$; $\chi^2_{(3)}=27.44, p<0.0001$.

[4] $\chi^2_{(2)}=27.23, p<0.0001$.

[5] $\chi^2_{(1)}=7.59, p<0.01$ and $\chi^2_{(3)}=27.44, p<0.0001$.

true for the less advantaged members of our sample: people with low education, low income, and rural origins. This finding supports that of several qualitative research studies conducted in China on the use of the mobile phones by migrant workers (Law and Peng, 2004a, b, 2005, 2006).

The number of personal contacts and social relationships has increased thanks to the mobile phone, but has there been a corresponding increase in the quality of these contacts and relationships? Did the mediation provided by mechanical means affect the quality of the communication? The answer, again, is clear: 68.6% of the respondents said that the use of the mobile phone made them feel closer to others, while almost one-third (31.4%) registered no change. This perception of closeness is significantly higher (80.0%) for respondents under 20 than for the respondents over 30 (45.5%).[6] For the generations who were practically born with mobile phones, the perception of feeling closer to others by means of this technology is strong. For those who encountered the mobile phone at a later point of their life, the perception that their communication patterns remain unchanged is higher. Furthermore, the perception of feeling closer to others is more widespread among those with rural backgrounds (76.2%).[7] For those respondents, the mobile phone clearly represents a way to maintain intimate relationships with their extended families. A large part of their emotional life is transmitted through mobile phones. Gender, education, and income do not have an effect on this perception of closeness.

These answers probably reflect the perception that mobile phones have enabled many contacts in everyday life that would have been impossible for the past. Still, one of the characteristics of the mobile phone is that users remain barricaded inside the limited network of people to whom they have given their mobile phone number. This negative aspect is reinforced by the fact that Chinese society is characterized by low trust.

To determine whether issue of trust limits the use of mobile phones, we asked, "Would you give your mobile number to someone you met only once or twice?" Almost half of the respondents (47.3%) answered positively to this question. This result suggests that mobile phone use is a new terrain, which has engendered a new social trust. It is worth noting that the respondents most likely to respond positively were between 21 and 30 (54.8%)[8] and urban (58.4%).[9] Gender did not play a role. The effects of education and income, however, were significant: 80.7% of the less educated respondents say that they would not give their mobile number to people they do not know well, while respondents with at least one degree are much more likely to give it (60.8%).[10] In terms of income bracket, 64.8% of the respondents with no income would give their number (the highest

[6] $\chi^2_{(3)} = 12.88, p < 0.01$.
[7] $\chi^2_{(1)} = 12.42, p < 0.0001$.
[8] $\chi^2_{(3)} = 15.35, p < 0.01$.
[9] $\chi^2_{(1)} = 23.10, p < 0.0001$.
[10] $\chi^2_{(2)} = 60.35, p < 0.0001$.

degree of trust), while 66.4% of those who earn about 1,000 *yuan* are reluctant to give their number.[11] These results seem to suggest that social experimentation, which involves developing new levels of trust, is more likely to be carried out by adults, urbanites, and well-educated people, on the one hand, or by people with no income who do not have anything to lose, on the other hand.

The question of social trust is a crucial one in China. The awareness that new technologies (especially the Internet but, to a lesser degree, the mobile phone) enlarge social capital/*guanxi* has obliged citizens to confront their issues with trust. Mobile phone users are an interesting litmus test; their behavior illustrates changes in social trust and the direction of those changes. When mobile networks were expanded to include Europe, youths called random numbers in order to meet new people. In China, where the mobile phone is often used also to chat on the web, we asked, "Would you give your mobile phone number to strangers, e.g., correspondents on the Internet?" The number of positive responses was less than half that of the earlier question. Only about one-fifth of our respondents (19.7%) said yes. The highest percentages were 23.2% and 24.5% of males and females, respectively, in the 21- to 25-year age group.[12] Education also played a role in the responses: those with a low level of education were more reluctant to give their telephone number (89.1%), while 23.5% of respondents with a high level of education were willing to disclose it.[13] When the issue of trust is probed, the identity of social experimenters becomes even more restrictive: they are male, well educated, and young adults.

If we want to adopt a more objective measure of the real dimension of the mobile connectivity, then we should look at the number of persons who are called by people with mobile phones. The average number of people called by our respondents was 17.3 (s.d. 31.9). In a study carried out in Japan, Miyata et al. (2005: 439) found that respondents had an average of 30.2 addresses in their telephones. In this study, income was the only factor that influenced the dimension of the communicative sphere mediated by the mobile phone: the higher the income, the greater the number of mobile contacts. [14]

The importance of one type of technology in everyday life can be measured by comparison with similar technologies. How did the respondents rank the mobile phone among ICT devices? The results are illustrated in Table 4.1.

Table 4.1 reveals that the respondents considered the mobile phone the most important technology, followed by the Internet and the computer. Furthermore, the judgment expressed about the mobile phone is compact and homogeneous among all the respondents, while that about the Internet and the computer is much more variable, as standard deviation indicates. This is also a proof of the large, transversal appreciation of the mobile phone among the whole sample.

---

[11] $\chi^2_{(3)} = 26.93$, $p < 0.0001$.
[12] $\chi^2_{(1)} = 5.10$, $p < 0.05$ and $\chi^2_{(3)} = 10.46$, $p < 0.05$.
[13] $\chi^2_{(2)} = 8.66$, $p < 0.05$.
[14] Univariate ANOVA with factor between income: $F_{(3,453)} = 3.19$, $p < 0.05$.

Table 4.1 Evaluation of the importance of ICT devices

Technology	Mean	Standard deviation
Mobile phone	4.35	0.94
Internet	4.28	1.28
Computer	4.24	1.31
Television	3.76	1.26
Fixed phone	3.53	1.23
Newspaper	3.08	1.28
Radio	2.50	1.26

The perceived importance of these technologies was measured using a 5-point scale, where *1* not important, *5* very important

Table 4.1 also shows that the respondents feel that new technologies have outdistanced the mass media and the fixed telephone. These findings reveal an attitude that is very different from the European perspective: in Europe, the mass media is deeply rooted in everyday life, and this shows limited signs of changing. The television is still considered the most important technology (Fortunati and Manganelli 1998).

In this study, women gave higher scores than men to all the technologies except the fixed and mobile phone, where their judgments were very close to men's.[15] Age also accounts for significant differences but only with respect to the assessments of the computer and the Internet, whose importance decreases as the age of respondents increases.[16] This is not a surprising result; it confirms that the computer and Internet are particularly attractive to young adults. Education also is an important variable when evaluating technologies.[17] The respondents' judgment was more likely to be favorable toward these four technologies, namely computer, the Internet, newspaper, and mobile phone, with an increase in their education. Income also plays a role, especially in the assessment of the importance of the computer, the Internet, and the mobile phone.[18] The respondents' scores were higher as their income rose. The middle and upper classes are probably able to incorporate technology more easily and more profitably into their daily life.

In deciding on the relative importance of these technologies, the urban/rural variable does not weigh significantly, except in the case of the computer and the Internet.[19] Respondents who were born in a city gave a significantly higher score to

[15] For TV: $t_{(485)}=-2.15, p<0.05$; for the radio: $t_{(482)}=-2.70, p<0.01$; for the computer: $t_{(481)}=-4.75, p<0.0001$; for the Internet: $t_{(480)}=-2.46, p<0.05$; for the newspaper: $t_{(484)}=-2.23, p<0.05$.

[16] Univariate ANOVA with factor between age: for the computer ($F_{(3,478)}=4.43, p<0.01$); for the Internet ($F_{(3,477)}=7.99, p<0.0001$).

[17] Univariate ANOVA with factor between education: for the computer ($F_{(2,479)}=94.55, p<0.0001$), for the Internet ($F_{(2,478)}=71.49, p<0.0001$), for the newspaper ($F_{(2,482)}=4.68, p<0.01$), and for the mobile phone ($F_{(2,481)}=3.70, p<0.05$).

[18] Univariate ANOVA with factor between income: for the computer ($F_{(3,477)}=21.93, p<0.0001$), for the Internet ($F_{(3,476)}=16.51, p<0.0001$), and for the mobile phone ($F_{(3,479)}=2.63, p<0.05$).

[19] For the computer: $t_{(478)}=7.93, p<0.0001$ and for the Internet: $t_{(477)}=7.81, p<0.0001$.

Table 4.2 Effects of demographic variables and perceptions on the judgments of importance of the mobile phone (stepwise multiple regression)

	B	β	t	$p<$
Perceived increased connectivity	0.23	0.19	4.09	0.0001
Age	0.02	0.13	2.72	0.01
Constant	2.93		9.77	0.0001
Adjusted R^2	0.05*			
N	448			

*$p<0.0001$

Table 4.3 Relative importance of the various purposes of use of the mobile phone

Purpose of use	Mean	Standard deviation
Social communication	4.36	1.12
Study/work	3.16	1.63
Amusement	2.96	1.56

both these technologies. The computer and the Internet seem to be more urban-oriented technologies because their dissemination and adoption relies on social learning processes (Williams et al. 2005) that are less accessible in rural areas.

But what are the most crucial variables in explaining the assessments of the importance of the mobile phone? To answer to this question, we ran a linear regression (see Table 4.2). We used certain sociodemographic variables as predictors, including gender (dummy variable 0=M, 1=F), urban/rural origin (dummy variable 0=city, 1=countryside), age (measured in years), education, and income. In addition, psychological variables related to the perception of connectivity, communicative sphere, and emotional distance, which had already been analyzed, were considered, as well as the respondent's number of usual contacts. The aim was to analyze the relative weight of these variables on the judgments of the importance of the mobile phone.

Two variables play significant roles in determining the importance of the mobile phone: perceived connectivity and age. The more respondents perceive that their connectivity is increased due to the mobile phone, the more they are likely to attribute a very important role to this device. This attribution also increases with age.

The purpose of one of the questions in our survey was to determine what users believed was the main purpose of mobile phones—amusement, social communication, or study/work. The response was measured on a 5-point scale, where 1=not important and 5=very important. The results are illustrated in Table 4.3.

Social communication was considered to be the main purpose of the mobile phone, followed by study/work, and, finally, amusement. There are significant differences in the assessments of the use of the mobile phone on the part of urban and rural respondents.[20] Respondents who were born in a village place a higher importance on amusement, while respondents who were born in a city attribute more

[20] For amusement ($t_{(473)}=-2.36$, $p<0.05$) and social communication ($t_{(473)}=4.15$, $p<0.0001$).

Table 4.4 Effects of demographic variables and perceptions on the use of the mobile phone for amusement purposes (stepwise multiple regression)

	B	*β*	*t*	*p* <
Education	−0.20	−0.23	−4.97	0.0001
Perceived augmented communication	0.41	0.13	2.83	0.01
Constant	3.04		6.09	0.0001
Adjusted R^2	0.08*			
N	448			

*$p < 0.0001$

importance to social communication. Age was a significant factor only in terms of the use of mobile phone for study/work purposes: the importance associated with this purpose varied according to the different ages.[21] As to education, respondents' judgment on the mobile phone use for amusement decreases in correspondence of the increase of the education level, while the importance given to social communication and study/work respondent is high for those with a basic education.[22]

Finally, income likewise originates significant differences: people with no income and people with more than 3,000 *yuan*[23] grant very little importance to the use of mobile phones for amusement. Respondents with low incomes gave the lowest score to the mobile phone use for social communication and study/work purposes.

We performed regression analysis to identify the variables influencing the judgments on the three uses we identified: the results for amusement are shown in Table 4.4, for social communication in Table 4.5, and for work/study in Table 4.6. The predictors are the same as those used in the previous analysis.

In the regression analysis for amusement, it emerged that the significant predictors are education and the perceived changes in one's sphere of communication. The higher the education level, the less likely that the mobile phone is used for amusement. For people with low levels of education, the mobile phone is an important resource for amusement; more educated people may have access to a wider range of sources of amusement. Furthermore, users who perceive that their social and communicative sphere is expanded thanks to the mobile phone are more likely to use the mobile phone for amusement.

The regression analysis for the use of the mobile phone for social communication purposes revealed that the two significant predictors are urban/rural origin and the perceived changes to the sphere of communication (see Table 4.5). Respondents

[21] Univariate ANOVA with factor between age: $F_{(3,473)} = 3.67, p < 0.05$.

[22] Univariate ANOVA with factor between education: for amusement ($F_{(2,474)} = 15.30, p < 0.0001$), for social communication ($F_{(2,474)} = 12.98, p < 0.0001$), and for study/work ($F_{(2,474)} = 4.59, p < 0.05$).

[23] Univariate ANOVA with factor between income: for amusement ($F_{(3,472)} = 3.41, p < 0.05$), for social communication ($F_{(3,472)} = 7.77, p < 0.0001$), and for study/work ($F_{(3,472)} = 5.78, p < 0.01$).

Table 4.5 Effect of demographic variables and perceptions on the use of the mobile phone for communicative purposes (stepwise multiple regression)

	B	*β*	*t*	*p*<
Urban/rural origin	−0.36	−0.17	−3.6	0.0001
Perceived augmented communication	0.27	0.13	2.68	0.01
Constant	3.89		15.10	0.0001
Adjusted R^2	0.03*			
N	448			

*$p<0.0001$

Table 4.6 Effect of demographic variables and perceptions on the use of the mobile phone for study or work purposes (stepwise multiple regression)

	B	*β*	*t*	*p*<
Age	0.03	0.12	2.26	0.05
Perceived augmented communication	0.60	0.19	4.04	0.0001
Income	0.09	0.13	2.37	0.05
Constant	0.473		0.929	n.s.
Adjusted R^2	0.06*			
N	448			

*$p<0.0001$

born in a rural area are less likely to use mobile phones for social communication. Migrants must overcome both the pain of separation and the difficulty of forging new relationships in a new environment, while urban people use their mobile phones to sustain and expand their stable social capital. Furthermore, evaluations of the importance of the mobile phone use for communicative purposes are positively influenced by the perception that these devices expand one's sphere of communication and enhance social and communication networks.

When the use of the mobile phone for work or study purposes (see Table 4.6) is analyzed, three significant predictors emerge: age, perceived changes in the sphere of communication, and income.

As age and income increase, an increased importance is given to the use of the mobile phone for study or work purposes. Furthermore, if one perceives that the sphere of communication is increased by the use of the mobile phone, one is more likely to consider the phone important for work or study purposes.

4 Conclusion

These results may aid in the development of a comprehensive sociological understanding of the adoption and use of mobile phones in contemporary China. According to our respondents, the mobile phone is the technology that has contributed the most to the reconfiguration of the communicative sphere during the last decade. Its use accompanied and supported the extraordinary process of Chinese modernization.

Our findings, which reinforce those of previous qualitative research, indicate that it is the less advantaged members of our sample (people with low levels of education, low income, and rural origins) who felt that the mobile phone widened their communicative and social networks not only quantitatively but qualitatively. This result is surprising as it refers to that part of our Beijing sample who, we assume, have experienced the social hardships described in the introduction: uprooting from the country, isolation during childhood because of the one-child policy, the impact of the new capitalist system with a large manufacturing sector, and the growth of a mass consumption and goods market. Our respondents seem to have been able not only to survive in these processes but also to increase their social capital (Ling et al. 2003; Fortunati 2007) and their traditional *guanxi*. The social empowerment that these people experience through the mobile phone is undeniable, and from this point of view, the mobile phone is a "technology of freedom" (de Sola Pool 1988) that, He (2008) argues, may even contribute to the expansion of democracy. At the same time, the mobile phone can do little to hinder the social stratification that is reshaping Chinese society. The cost of its use is almost prohibitive for those who rely on it the most; these people must spend a large part of their salary to pay for the device and its use. Of course, they are obliged to limit their personal mobile calls for economic reasons. Still, their mobile expenses remain very high and contribute to their inability to escape the lowest strata of the Chinese population. This is an ironic situation, which inspired Law (unpublished manuscript) to liken the mobile phone to an empty shell, representing but denying modernization to the lowest social strata. After all, as Ferrarotti (1970: 114) observes, "technology might increase production and accelerate development, but it does not include in itself the tools to resolve the problems related to development. Technology is able to offer means but is not able to indicate the aims."

The fact that in China the social appropriation of this technology has extended to the most disadvantaged strata seems to indicate that the mobile phone responds to a very fundamental need in the whole population. It is likely that Chinese people are drawn to the mobile phone, and especially the SMS, because it is a means of overcoming the rigidity and the emotional censures connected with body-to-body encounters. In this respect, China differs from Europe, where these social strata came into the market much faster than those did in Europe. Age also plays a less important role in China than it does in Europe. Although the women in our study gave higher scores to all the technologies (except for fixed and mobile telephones) than the men, women seem to have more difficulties than men in using the mobile phone to enhance their sphere of communication. Probably, limitations associated with the traditional lives of women discouraged women from receiving the same benefits from mobile phones as men.

Another interesting finding of the study is the difference in the role played by the mobile phone and the role played by the Internet. The former is much more socially inclusive than the latter. While mobile phones were owned by even the poorest respondents, Internet use was mostly confined to the middle and upper classes.

This latter part of Chinese population is assigning to the Internet the role of helping to overcome the informational/communicative isolation of the country from the rest of the world. The computer/Internet, in addition to the means of transport, especially the aeroplane, is contributing in an important way, in the context of the rapid industrialization, to the integration of China in the world's economic and political system. Furthermore, the mobile phone and the Internet receive the highest appreciation by the respondents, in contrast to old technologies, such as TV, radio, and the newspaper. This suggests that the new developed countries are receptive to communication and information technologies and privilege the new media.

Finally, the regression analyses show that demographic variables affect these judgments in sporadic ways. The one variable significant in every analysis is the user's perception of increasing the sphere of communication. This variable was crucial in determining the assessment of the importance of the mobile phone among our respondents.

As we already underlined, this research has the limit that the data presented and discussed here are obtained from a convenience sample. However, although it is not possible to generalize the results to the population of reference, this study might be a good starting point for future research.

References

Chu, W.-C., & Yang, S. (2006). Mobile phones and new migrant workers in a south China village: An initial analysis of the interplay between the "social" and the "technological". In P. Law, L. Fortunati, & S. Yang (Eds.), *New technologies in global societies* (pp. 221–244). Singapore: World Scientific Publisher.

de Sola Pool, I. (1988). *Technologies of freedom*. Cambridge: Cambridge University Press.

Ferrarotti, F. (1970). *Macchina e uomo nella società industriale*. Roma: ERI.

Fortunati, L. (2007). Immaterial labor and its machinization. *Ephemera: Theory & Politics in Organization, 7*(1), 139–157.

Fortunati, L., & Manganelli, A. (1998). La comunicazione tecnologica: Comportamenti, opinioni ed emozioni degli Europei. In L. Fortunati (Ed.), *Telecomunicando in Europa* (pp. 125–188). Milano: Angeli.

Fortunati, L., & Yang, S. (2008). The identity and sociability of the mobile phone in China. In J. M. Aguado & I. J. Martinez (Eds.), *Sociedad Móviles: Tecnología, identidad y cultura* (pp. 137–165). Madrid: Biblioteca Nueva.

Fortunati, L., Manganelli, A., Law, P., & Yang, S. (2008). The 'Mobile' face of contemporary China. *Telektronikk, 104*(2), 57–67.

Fortunati, L., Manganelli, A., Law, P., & Yang, S. (2009). Practices of mobile phone use in Beijing. In L. Stillman, G. Johanson, & R. French (Eds.), *Communities in action: Papers in community informatics* (pp. 163–181). Newcastle: Cambridge Scholars Publishing.

Fortunati, L., Manganelli, A., Law, P., & Yang, S. (2010). The mobile phone use in Mainland China: Some insights from an exploratory study in Beijing. *Telematics and Informatics, 27*(4), 404–417.

He, Z. (2008). SMS in China: A major carrier of the nonofficial discourse universe. *The Information Society, 24*(3), 182–190.

Law, P., & Peng, Y. (2004a, June). *Cellphone, internet and the SARS epidemic*. Paper presented at the international conference on Mobile Technologies and Health: Positive and Negative Aspects, Udine, Italy.

Law, P., & Peng, Y. (2004b, October). *The use of cellphones among migrant workers in Southern China*. Paper presented at the international conference on Communication and Social Change, Seoul, Korea.
Law, P., & Peng, Y. (2005, October). *Cellphones and the social lives of migrant workers in southern China*. Paper presented at the international conference on Mobile Communication and Asian Modernities II: Information, Communications Tools & Social Changes in Asia, Beijing, PRC.
Law, P., & Peng, Y. (2006). The use of mobile phones among migrant workers in southern China. In P. Law, L. Fortunati, & S. Yang (Eds.), *New technologies in global societies* (pp. 245–258). Singapore: World Scientific Publisher.
Ling, R., Yittri, B., Anderson, B., & DiDuca, D. (2003). Mobile communication and social capital in Europe. In K. Nyíri (Ed.), *Mobile democracy: Essays on society, self and politics* (pp. 359–374). Wien: Passagen Verlag.
Miyata, K., Wellman, B., & Boase, J. (2005). The wired – and wireless – Japanese: Webphones, PCs and social networks. In R. Ling & P. E. Pedersen (Eds.), *Mobile communications: Re-negotiation of the social sphere* (pp. 427–449). London: Springer.
Tomba, L. (2003). L'ascesa della classe media. *Aspenia, 23*, 95–100.
Weber, M. (2003). Alla ricerca di alta tecnologia. *Aspenia, 23*, 101–106.
Williams, R., Stewart, J., & Slack, R. (2005). *Social learning in technological innovation*. Cheltenham: Edward Elgar.
Yang, B. (2005, October). *Social networks and individualism: The role of the mobile phone in China's social transformation*. Paper presented at the international conference on Mobile Communication and Asian Modernities II: Information, Communications Tools & Social Changes in Asia, Beijing, PRC.
Yuen, S. P., Law, P., & Ho, Y. Y. (2004). *Marriage, gender, and sex in a contemporary Chinese society*. New York: M.E. Sharpe.

Chapter 5
The "Mobile" Face of Contemporary China

Leopoldina Fortunati, Anna Maria Manganelli, Pui-lam Law, and Shanhua Yang

1 China's Position at the Crossroads of Industrialization, Technology, and Modernity

Sombart (1911) argued that each culture adapts technology to the context of its daily life. But what does this mean in the case of China, a country with a culture that differs strikingly from its Western counterpart? How is new technology such as the mobile phone appropriated and domesticated in China? The effects of technological advances are widely studied in the West, but this research is not necessarily applicable to China. Why did the mobile phone, in particular, become so immediately and widely popular among the Chinese? Why is a mobile phone so attractive that Chinese men and women are willing to spend 1–3 months' wages to own one?

With almost 500 million mobile phones, China is now the primary market for this technology and has become one of the world's leading ICT nations. China is a recently industrialized country; its rate of development is very high in comparison to countries where industrialization is well established. It is a country where the diffusion, use, and appropriation of technology, especially information and communication technology, play a strategic role in the development of production systems (Weber 2003). Industrialization and technology are inextricably bound with

L. Fortunati
Faculty of Educational Sciences, University of Udine, Udine, Italy

A.M. Manganelli
Department of General Psychology, University of Padova, Padova, Italy

P.-l. Law (✉)
Department of Applied Social Sciences, The Hong Kong Polytechnic University, Hong Kong, China
e-mail: sspllaw@inet.polyu.edu.hk

S. Yang
Department of Sociology, Peking University, Beijing, China

P.-l. Law (ed.), *New Connectivities in China: Virtual, Actual and Local Interactions*,
DOI 10.1007/978-94-007-3910-9_5, © Springer Science+Business Media B.V. 2012

another element: modernity (Ferrarotti 1970; Giddens 1991a, b). By "modernity," we refer to the radical changes these new developments effect in social structures, the configuration of social relationships, and the perception of self, which result in the reshaping of the material and immaterial organization of the domestic sphere and civil society. Generally, the success of a technological device depends on its ability to make people believe that it will help them to deal with these changes. Technology and modernity are, therefore, closely interconnected.

In this chapter we address other research questions, especially those related to the relational sphere of the mobile phone and to users' attitudes towards the device. On the basis of questionnaires personally administered to a convenience sample of 487 respondents, this study investigates the following areas: the identity of the main recipient of calls, the frequency of mobile communicative exchanges, the type of contact (calls or SMS), the nature of SMS content, the respondents' attitudes towards the increasing complexity of the mobile phone, the respondents' passive or active approach to mobile communication, the advantages and disadvantages of the mobile phone, and possible substitutions for the device. Of course, a convenience sample of 487 respondents can hardly provide a basis for generalizations about Chinese population as a whole. However, the results of the research suggest the most important patterns of the mobile phone use, which should be examined in further research. Thus, the data presented here is intended to provide a starting point for a more in-depth inquiry into various aspects of the mobile phone use in China.

2 Aim and Method

The purpose of the study is broad and covers a number of aspects of the appropriation and domestication of the mobile phone in China. In this chapter, however, we will confine ourselves to the specific issues identified in the previous section. As mentioned in Chap. 4, the data were collected during personally administered surveys conducted in March 2006 in Beijing. The demographic variables included in the research were gender, age, education, income, and place of origin (rural or urban). The participants ($n=487$) were Beijing inhabitants: 280 males (57.5%) and 207 females (42.5%). The respondents' ages varied from 15 to 60, but the large majority of the sample (86%) were under 30. Almost half were born in the countryside. Nearly a quarter had only a basic education (secondary school); one-fifth have a mid-level education (high school or some years of college); and more than half had at least one degree. With respect to the income, 21.6% had no income; 30.6% earned 1,000 *yuan* or more per month; 34.3%, from 1,001 to 4,000 *yuan*; and 13.1%, 4,000 *yuan*.[1]

[1] One euro is about 10 *yuan*.

3 Results

People who use the mobile phone as a means of conducting their interpersonal relationships end up reshaping their social sphere. Since the phone allows users to be highly selective, they are able to shape the relational sphere a rigid and self-determined way. Therefore, the first question of the survey was a follows: "Which interpersonal relationships are included in the social sphere maintained by your mobile phone?" We know that not all interpersonal relationships involve the same degree of mobile phone use. There are relationships that encourage only sporadic mobile phone use, and others that rely on it heavily. Table 5.1 illustrates this situation.

This table provides several insights into the structure of relational spheres in Beijing today. First, the mobile is not primarily used in the sphere of family relationships. Within families, it is more often used as a means of communication between husbands and wives than between parents and children (21.8% and 15%, respectively). It acts more horizontally than vertically. This finding is supported by the fact that the mobile phone used primarily to sustain friendships and keep in touch with schoolmates (40.9%). Colleagues (12.3%) are in fourth place as recipients of mobile calls. Mobile phones are very seldom used to contact people from one's hometown (though a relationship with one's birthplace has been historically very important for Chinese people). Only two respondents indicated that they called inhabitants of their hometown on their mobile phone. Probably the costs due to distance, in addition to technological annoyances and other psychological and communicative factors, discourage the use the mobile phone with hometown acquaintances.

In the organization of their mobile communicative sphere, men and women differ only insofar as the former are more inclined than the latter to contact a colleague by means of a mobile phone (18.4% and 6.6%, respectively).[2] Age also plays a part in mobile phone use: those who are between 26 and 30 with a mid-level education are more apt to contact their colleagues by means of this device, while those under

Table 5.1 Most frequent recipients of mobile phone calls

Recipient	Frequency	Percentage
Your friends or schoolmates	199	40.9
Your partner	106	21.8
Parents, children, or other relatives	73	15.0
Your colleagues or other working partners	60	12.3
Persons from your hometown	2	0.4
Others	11	2.2
NA	36	7.4
Total	487	100.0

[2] Gender: $\chi^2_{(4)} = 16.24, p < 0.01$; age: $\chi^2_{(12)} = 83.64, p < 0.0001$; education: $\chi^2_{(8)} = 59.08, p < 0.0001$; $\chi^2_{(12)} = 77.47, p < 0.0001$; geographical origin: $\chi^2_{(4)} = 10.36, p < 0.05$.

Table 5.2 Average frequency of contacts with partner, parents/children, friends, colleagues

	Partner	Parents, children, or other relatives	Friends or schoolmates	Colleagues
Average frequency	3.95	2.69	3.02	3.02

The 5-point scale was as follows: *1* not often, *2* once a week, *3* many times in a week, *4* once a day, *5* several times a day

25 years are the most likely to use it to contact their friends or schoolmates. This suggests that the former group is obliged to devote more time to work problems and relationships with colleagues. For this group in particular, the mobile phone is a crucial work tool. Well-educated respondents follow another pattern; they are more inclined to use the mobile to work out their interpersonal relationships. In fact, these respondents make half of their mobile calls to friends and a third to their partners. Other patterns are created by income level: 68.8% of those with no income contact friends, while people with an income above 1,000 *yuan* are more likely to call their colleagues. Finally, with respect to geographical origin, it appears that urbanites are more likely to nurture sentimental relationships by means of the mobile phone than those with a rural background. The majority of respondents who contact their partner via the mobile phone came from the city. The relational spheres of the respondents, as shown by the frequency of their contacts with friends, partners, relatives, and colleagues, are illustrated in Table 5.2.

It is worth noting that our respondents use the mobile with varying frequency depending on the nature of their relationship with the recipient. The lowest frequency of calls is evident in the relationship between parents and children; the mean 2.69 indicates an obligatory call once a week. This is the relationship where the mobile phone plays the least critical role because when parents and children no longer live together, they need to have a regular, but not too frequent, contact to reassure one another that they are doing well. Partners are the people that our respondents contact most intensively, almost once a day ($M = 3.95$). There are many things that couples need to regulate and coordinate; more importantly, the calls give them an opportunity to express their feelings and experiences. Contact with friends or schoolmates and contact with colleagues were equally frequent ($M = 3.02$, i.e., many times a week). The survey indicated that the mobile phone acts as an important means of conducting work relationships. Not surprisingly, we found differences based on gender: women are more frequent callers of partners, parents/children, and friends/schoolmates, while men show a greater tendency to call colleagues.[3] In China, as in the West, there is a gender division of labour regarding the maintenance of social relationships. Maintaining social bonds is a woman's task, part of the domestic sphere, while cultivating work relationships is more likely to be undertaken by men. There are slight differences based on geographical origin: those with urban backgrounds made more

[3] For partners: $t_{(305)} = -3.09, p < 0.01$; parents/children: $t_{(458)} = -3.470, p < 0.01$; friends/schoolmates: $t_{(454)} = -2.38, p < 0.05$; colleagues: $t_{(401)} = 2.56, p < 0.05$.

calls to call partners, parents/children, and friends/schoolmates, but they called colleagues at the same rate as respondents born in the country.[4] It seems that it is especially the domestic sphere to be more solicited on a communication level by the urban culture. The only significant differences related to age were in the frequency with which our respondents call friends/schoolmates and colleagues.[5] Not surprisingly, young respondents (25 and under) have more frequent communications with friends and schoolmates, while more mature respondents are more likely to call their colleagues. Education influences only the frequency with which respondents contact partners and friends/schoolmates: the higher the educational level, the more likely it is that there are frequent calls to partners and friends.[6] Income plays a role too: people with higher incomes tend to have a wider variety of recipients.[7] These results mean that the possibility to keep alive personal networks does not pass only through a personal, psychological ability, but it is connected also to these socio-demographic variables.

Another significant finding, which sheds light on the relational spheres of the respondents, is the number of people who our respondents contact by means of the mobile phone. The average for the sample is 17.32 (s.d. 31.81) per person. The men had more contacts than the women ($M = 20.36$ and 13.32, respectively).[8] There were no differences between respondents born in the countryside and those born in city. Among the other socio-demographic variables, only education and income are significant factors[9]: people with a mid-level education contact more people ($M = 27.46$) than those with other levels of education, and people with high or middle incomes have the largest number of mobile phone contacts ($M = 21.40$ and 22.23, respectively).

Let us look now at the means of contact, since the form of communication (calling or sending an SMS) provides insight into the content of various social relationships (see Table 5.3).

This table shows that each type of relationship is associated with a certain means of mobile contact. With parents or children, it is almost always a phone call. Parents and children communicate by means of the voice, even when relying on technology. Recourse to SMS is very limited, and in less than 10% of cases, the decision to use one means or the other depends on the context or situation. Friends and

---

[4] For partners: $t_{(303)} = 3.64, p < 0.0001$; parents/children: $t_{(455)} = 4.67, p < 0.0001$; friends/schoolmates: $t_{(451)} = 3.49, p < 0.01$.

[5] Univariate ANOVA with factor between frequency of contact with friends/schoolmates: $F_{(3,451)} = 25.87, p < 0.0001$; with colleagues: $F_{(3,398)} = 3.73, p < 0.05$.

[6] Univariate ANOVA with factor between frequency of contact with partner: $F_{(2,303)} = 10.54$, $p < 0.0001$; with friends/schoolmates: $F_{(2,452)} = 14.65, p < 0.0001$.

[7] Univariate ANOVA with factor between frequency of contact with partner: $F_{(3.301)} = 7.67$, $p < 0.0001$; with parents/children $F_{(3,454)} = 4.02$, $p < 0.01$; with friends/schoolmates: $F_{(3,450)} = 16.71$, $p < 0.0001$; with colleagues: $\mathrm{F}_{(3,397)} = 5.89$, $p < 0.01$.

[8] Gender: $t_{(457)} = 2.36, p < 0.05$.

[9] Univariate ANOVA with factor between education: $F_{(2,455)} = 6.16, p < 0.01$ and income: $F_{(3,453)} = 3.18$, $p < 0.05$.

Table 5.3 The means of contact

Means	Partner	Parents, children, or other relatives	Friends and schoolmates	Colleagues
Calling	126	368	80	163
	25.9%	75.6%	16.4%	33.5%
Sending an SMS	131	41	300	109
	26.9%	8.4%	61.6%	22.4%
Varies according to context and situations	55	46	73	134
	11.3%	9.4%	15.0%	27.5%
NA	175	32	34	81
	35.9%	6.5%	6.9%	16.6%

This table is built with the results of several contingency tables

schoolmates communicate mainly through SMS and only sometimes by calling. The choice of means in these relationships is more sensitive to different contexts and situations (15%). The relationship with partners inhabits an area between those with parents/children and those with friends. In the case of partners, both means of communication—calling and sending SMS—are used in equal measure. Also, the effect of context and situation on the means of communication is not as limited as it is for parents/children and not as significant as it is for friends. Finally, contact with colleagues is more likely to be a call than an SMS, and the effect of context and situation is much more crucial (27.5%). Vertical relationships, that is, relationships based on a difference of power, such as those between parents and children and between different levels of personnel, seem to require the spoken word, while horizontal relationships, such as those between friends and partners, encourage uninhibited writing.

We will examine each type of social relationship, in turn, to determine the weight of each socio-demographic variable in each framework. In relationships with partners, the means of contact by mobile phone varies, revealing different styles of communication.[10] While men's behaviour is not different from women's, age is relevant. More than a third of the respondents under 20 use SMS to communicate with their partners, whereas 59.7% of those between 31 and 60 years make calls. Also, education shapes the means of communicating with a partner: 38.4% of respondents with a low level of education rely on calls, while most well-educated respondents (72.7%) exercise the prerogative to make a call or send an SMS according to different contexts and situations. Income also affects the likelihood of calling: respondents who gain between 1,000 and 3,000 *yuan* significantly more than the others do so. In relationships with parents/children, men are more likely than women to call their parents or children rather than sending an SMS (83.7% and 75.0%, respectively).[11] Those who earn less than 1,000 *yuan* are more

[10] Age: $\chi^2_{(6)}=36.50, p<0.0001$; education: $\chi^2_{(4)}=15.20, p<0.01$; income: $\chi^2_{(6)}=14.45, p<0.05$.
[11] Gender: $\chi^2_{(2)}=7.61, p<0.05$; income: $\chi^2_{(6)}=14.48, p<0.05$; geographical origin: $\chi^2_{(2)}=8.25, p<0.05$.

likely to call parents/children (87%). Respondents who were raised in the city are more likely to send an SMS than those from the country (12.6% and 5.3%, respectively), which leads to the corollary that those with rural backgrounds are more likely to call than urbanites (84.5% and 75.8%, respectively). In relationships with friends/schoolmates, men are more likely than women to call (24.6% and 7.6%, respectively); women are more apt to send an SMS and to adapt the means of communication to circumstances and events.[12] One-third of respondents between 31 and 60 years call their friends, while more than 70% of respondents under 25 rely on SMS. Almost one-third of the respondents who earn over 1,000 *yuan* make calls, while 82.2% of those with no income use SMS. Finally, in relationships with colleagues, men are much more likely than women to make a call.[13] Age introduces a dichotomy between young (under 25) and more mature (26–60) respondents: the former group is likely to send an SMS, the latter to make calls. More than half of respondents with a mid-level education are likely to use their mobile phone to call their colleagues, while more than a third of interviewees with a low level of education prefer to send an SMS. Among those who vary means of contact on the basis of context and situation, 61.9% are people with higher levels of education. Income also played a significant role in the responses. At least half of those with an income over 1,000 *yuan* prefer to call, while a third of respondents with no income adapt the means according to the circumstances. Respondents with urban backgrounds are more likely than those with rural backgrounds to vary the means according to the circumstances (32.9% and 26.3%, respectively). Respondents born in the country prefer to send an SMS in 30.4% of cases, but this is true of only 18.2% of urbanites.

Another very important element to consider when determining the communication profile of a country and its organization of interpersonal relationships is people's attitude towards the two aspects of mobile communication: giving and receiving calls. Unlike face-to-face communication, mediated communication requires cooperation between these two elements of the process since the agents are separated by space and, in some instances, time as well. Of course, there are people who prefer to call and people who prefer to be called. These attitudes may be described as active or passive approaches to communication. However, the analysis of the passive/active attitude towards communication cannot simply be understood in a psychological or socio-demographic framework. Understanding the reasons for these patterns (in calling and being called) is complicated by the fact that communication is subjected to the social organization of power. In some situations, those with greater power assume the initiative to call and talk, and in others, they prefer to receive communications initiated by others. Furthermore, the organization of power leads to different patterns and rituality depending on the specific culture. In Beijing, we found that the percentage

---

[12] Gender: $\chi^2_{(2)}=25.79$, $p<0.0001$; age: $\chi^2_{(6)}=39.35$, $p<0.0001$; income: $\chi^2_{(6)}=39.34$, $p<0.0001$.

[13] Gender: $\chi^2_{(2)}=16.19$, $p<0.001$; age: $\chi^2_{(6)}=51.96$, $p<0.0001$; education: $\chi^2_{(4)}=19.18$, $p<0.001$; income: $\chi^2_{(6)}=45.63$, $p<0.000$; geographical origin: $\chi^2_{(2)}=8.09$, $p<0.05$.

Table 5.4 Giving or receiving calls: prevalent use of the mobile phone by gender

	Gender		
Type of use	Males	Females	Total
Giving a call	93	85	178
	35.4%	42.7%	38.5%
Receiving a call	70	31	101
	26.6%	15.6%	21.9%
Nearly half and half	100	83	183
	38.0%	41.7%	39.6%
Total	263	199	462

In this table, 25 N.A. are not reported

of respondents who claim that they usually give calls is higher compared to that of respondents who say that they usually receive calls (see Table 5.4).

So it appears our Chinese respondents see themselves as more active than passive in mobile communication. Considering this aspect from the perspective of gender allows us to understand how important ritual aspects of communication are organized in a country.[14] In this survey, for example, there is a significant difference between men and women: women are more likely to take the initiative and make a call, opening up the communicative process. Men are more frequently the recipients of calls. This result may lead us to hypothesize that in Chinese culture power is identified with the capacity to attract communications from others, almost as a form of homage. But this hypothesis is weakened when we consider the effects of education. Respondents with a mid- and high-level education give more calls than they receive. Also, respondents with urban backgrounds are more inclined to initiate a call than those with rural backgrounds (46.8% and 30.1%, respectively). These findings suggest that Chinese are proactive in communication and, on the whole, are more likely to take the communicative initiative than to await incoming calls that are perceived as a form of homage. The fact that women's calls are mainly related to interpersonal relationships in the domestic sphere provides additional support to the theory that social communication is part of their domestic role.

But what are the most crucial variables when it comes to accounting for the varying frequency of mobile calls to family, friends, and colleagues? To answer this question, we ran a linear regression analysis: the results for mobile calls to a partner are shown in Table 5.5, to parents/children and other relatives in Table 5.6, to friends/schoolmates in Table 5.7, and to colleagues in Table 5.8. We used some socio-demographic variables as predictors, including gender (dummy variable 0=M, 1=F), urban/rural origin (dummy variable 0=city, 1=countryside), age (measured in years), education, and income. We also included variables related to mobile phone practice, such as familiarity with the mobile phone (in years), quantity of calls made and messages sent with the mobile phone, and the range of social contacts maintained by this device.[15]

[14] Gender: $\chi^2_{(2)}=8.39, p<0.05$; education: $\chi^2_{(4)}=10.35, p<0.05$; geographical origin: $\chi^2_{(2)}=14.05, p<0.0001$.

[15] We illustrated the results related to these variables in Fortunati et al. (2009).

Table 5.5 Results of multiple regression analysis for partners (stepwise method)

	B	*β*	*t*	*p*<
Constant	3.53		6.09	0.0001
Education	0.09	0.12	2.00	0.05
Average number of SMS sent per day	0.01	0.17	3.02	0.01
Income	0.16	0.26	3.92	0.0001
Age	−0.04	−0.19	−3.00	0.01
Gender	0.36	0.13	2.28	0.05
Adjusted R^2	0.14*			
N	285			

$^*p<0.0001$

Table 5.6 Results of multiple regression analysis for parents/children and other relatives (stepwise method)

	B	*β*	*t*	*p*<
Constant	1.95		7.28	.0001
Geographical origin	−0.36	−0.16	−3.30	0.01
Average number of SMS sent per day	0.01	0.13	2.68	0.01
Age	0.03	0.14	2.86	0.01
Gender	0.28	0.12	2.54	0.05
Adjusted R^2	0.07*			
N	426			

$^*p<0.0001$

The regression analysis for the frequency of mobile contacts with partners revealed five significant predictors: four were socio-demographic and one was related to mobile phone practices (see Table 5.5). The main predictor was income. The number of contacts with partners increases with higher incomes. This is not surprising; affluent respondents can afford a higher frequency of communication because they can easily pay mobile fees. The frequency of contact increases as age decreases. This inverse relationship is also easily understandable: young people are at the stage of life when long-term relationships are established, so they dedicate more effort to communication. Furthermore, frequency of contacts is positively influenced by the practice of sending SMS. The more respondents are inclined to send an SMS, the more likely they are to contact their partner regularly. Gender also plays a role: women are more likely to contact their partner than men. In China, as in the West, the responsibility to nurture sentimental relationships, to coordinate daily activities, and to establish a domestic framework is shouldered primarily by women (Yuen et al. 2004). Finally, respondents with a high level of education are more likely to use the mobile phone to contact their partner. The cultivation of sentimental relationships through conversation is one of the consequences of a high level of education.

In the regression analysis for mobile contacts with parents/children or other relatives, the significant predictors were found to be geographical origin, gender, age, and the average number of SMS sent (see Table 5.6). Respondents born in the

Table 5.7 Results of multiple regression analysis for friends/schoolmates (stepwise method)

	B	β	t	$p<$
Constant	3.69		9.41	0.0001
Age	−0.05	−0.21	−3.72	0.0001
Education	0.15	0.18	4.05	0.0001
Average number of SMS sent per day	0.01	0.16	3.47	0.01
Income	−0.11	−0.17	−3.17	0.01
Adjusted R^2	0.17*			
N	422			

*$p<0.0001$

country were less inclined to contact parents/children than those born in the city. Internal migration makes it more difficult to sustain family relationships due to the cost and to the fact that vertical communication tends to become more strained and, therefore, less frequent. In addition, as Law and Peng (2007) have observed, there have been changes in lifestyle among recent generations of immigrants, who are more likely to be consumers and less likely to send their money home to take care of their parents and children. The number of contacts increases with age. Youths wait for parents to initiate contact; they are eager to cut the umbilical cord that keeps them united to their families and, for this reason, they tend call their parents less frequently. In contrast, parents try to sustain their relationship with their children and exert some control over their behaviour. The frequency of parent/child contacts is increased if the parties use SMS. The more inclined the respondents are to send an SMS, the more likely they are to contact their parents/children regularly. The SMS is, in this context, the ideal means to maintain a relationship: it provides communication but at a distance. Women are more frequent communicators than men, both as daughters and as mothers.

In the regression analysis for mobile contacts with friends/schoolmates, the significant predictors were age, education, income, and the average number of SMS sent (see Table 5.7). With an increase in age, the likelihood of mobile contact decreased. This finding is not unexpected; as we age, we acquire more commitments and family/work responsibilities. This means we have less time to dedicate to friendships. The use of the mobile phone offers a very good picture of this phenomenon. Education is a powerful ally of friendship: it offers reasons, motivations, and means to continue to stay in touch with friends. Education teaches the value of friendship, and the time we spend in school allows us to build friendships that may last for a lifetime. The higher the level of education, the more mobile contacts with friends and schoolmates. There are also correlations with income. But here we had an unexpected result. Research conducted in the West shows that, in order to maintain friendships, it is necessary to invest money in means to stay in touch and meet (Allan 1979). Money is required to access, maintain, and use mobile phones, computers, the Internet, and even the fixed telephone; it is also required for evenings at restaurants or cinemas, etc. However, our research suggested the opposite. The less the respondents earn, the more likely they were to stay in contact with their friends

Table 5.8 Results of multiple regression analysis for colleagues (stepwise method)

	B	β	t	$p<$
Constant	2.19		14.24	0.0001
Average number of calls made per day	0.04	0.25	4.84	0.0001
Familiarity with the device	0.10	0.18	3.63	0.0001
Largeness of the relational sphere	0.01	0.10	2.06	0.05
Adjusted R^2	0.15*			
N	373			

$^*p<0.0001$

and schoolmates. One reason may be that the mobile phone serves a very important function for the poorest respondents—it is a generator of employment information. Also, the frequency of friends/schoolmates contacts is positively influenced by SMS use. The more respondents use SMS, the more they stay in contact with their friends/schoolmates. The SMS is in this context the ideal means to maintain a relationship without spending too much.

Finally, in the regression analysis for mobile contacts with colleagues, the significant predictors are the number of mobile calls made, familiarity with the mobile phone, and the largeness of the relational sphere maintained by this device (see Table 5.8). The predictors that explain the frequency of contact with colleagues are completely different from the predictors in other spheres. Here it is the number of calls that matters, not the frequency of SMS. Work requires a level of discretion that often makes written communication inadvisable. Such communication must be transitory in order not to leave traces; it needs a voice to express all the nuances necessary to deal with various power relationships. Furthermore, familiarity with the device is important, as people cannot risk embarrassing themselves with colleagues due an inadequate knowledge of mobile phone use. The longer that respondents had owned the device, the more intensively they used it to contact their colleagues. People who are familiar with the device demonstrate to their colleagues that they can manage technology efficiently. Finally, the largeness of the relational sphere plays a significant role. This variable acts as an index of empowerment: people who have a large number of contacts seem well connected.

4 Conclusion

In this chapter we examined the structure and the characteristics of mobile phone behaviour among our Beijing respondents. Many patterns are similar to those found in the West. But in at least two respects, the Chinese experience seems to be different. Our respondents in Beijing claim more contact with colleagues (Fortunati and Manganelli 1998), which suggests that, for the Chinese, the mobile phone is perceived as a work tool as well as a social device. Also, in China, the respondents who earned the least were the most likely to stay in contact with their friends and schoolmates via the mobile phone; this suggests that the device operates as a generator of social solidarity among peers and is, consequently, or particular value to the poorest respondents. This is in contrast to the situation in the West, where friendships generally are a "rare good" among people with low incomes (Allan 1979; Fortunati 1995). But this means also that probably it is from friendships, that is, from strong ties, that comes the so precious employment information. And this again is different from the Western experience, where, as Granovetter (1973, 1974) argues, are the weakest ties the most involved in helping to get a job.

There were not very significant differences found between men and women in this study, except that women's social contacts are more likely to relate to the domestic environment and the men's to the work world. And this lack of differences between men and women was unexpected in a country where power differences between men and women are still strong (Yuen et al. 2004).

Finally, modernity in China has been achieved by the urbanization of millions of peasants. The development of the coastal cities has been achieved at high personal and social costs, which have been borne primarily by those with rural backgrounds (Chu and Yang 2006). In addition to being uprooted from their home region, they encounter serious difficulties in the attempt to enter their new social sphere. In some respects, the mobile phone supports their efforts to overcome these social, economic, and cultural difficulties, but it does not serve to maintain their relationship with their place of origin.

References

Allan, G. A. (1979). *A sociology of friendship and kinship*. London: Allen & Unwin.

Chu, W.-C., & Yang, S. (2006). Mobile phones and new migrant workers in a south China village: An initial analysis of the interplay between the "social" and the "technological". In P. Law, L. Fortunati, & S. Yang (Eds.), *New technologies in global societies* (pp. 221–244). Singapore: World Scientific Publisher.

Ferrarotti, F. (1970). *Macchina e uomo nella società industriale*. Roma: ERI.

Fortunati, L. (Ed.). (1995). *Gli italiani al telefono* [Italians on the phone]. Milano: Angeli.

Fortunati, L., & Manganelli, A. (1998). La comunicazione tecnologica: Comportamenti, opinioni ed emozioni degli Europei. In L. Fortunati (Ed.), *Telecomunicando in Europa* (pp. 125–188). Milano: Angeli.

Fortunati, L., Manganelli, A., Law, P., & Yang, S. (2009). Practices of mobile phone use in Beijing. In L. Stillman, G. Johanson, & R. French (Eds.), *Communities in action: Papers in community informatics* (pp. 163–181). Newcastle: Cambridge Scholars Publishing.

Giddens, A. (1991a). *The consequences of modernity*. Cambridge: Polity Press.

Giddens, A. (1991b). *Modernity and self-identity: Self and society in the late modern age*. Cambridge: Polity Press.

Granovetter, M. (1973). The strength of weak ties. *American Journal of Sociology, 78*(6), 1360–1380.

Granovetter, M. (1974). *Getting a job: A study of contacts and careers*. Cambridge: Harvard University Press.

Law, P., & Peng, Y. (2007). Cellphones and the social lives of migrant workers in southern China. In R. Pertierra (Ed.), *The social construction and usage of communication technologies: Asian and European experiences* (pp. 126–142). Manila: The University of the Philippines Press.
Sombart, W. (1911). Technick un Kultur. *Archiv für Sozialwissenschaft und Sozialpolitik*, 33, 305–347 (tr.it. 1991, Tecnica e cultura, in T. Maldonado, ed. Tecnica e cultura. Il dibattito tedesco fra Bismark e Weimar, Milano: Feltrinelli, pp. 137–170).
Weber, M. (2003). Alla ricerca di alta tecnologia. *Aspenia, 23*, 101–106.
Yuen, S. P., Law, P., & Ho, Y. Y. (2004). *Marriage, gender, and sex in a contemporary Chinese society*. New York: M.E. Sharpe.

Chapter 6
A Preliminary Study on the Mobile Phone Use of Migrant Workers in Beijing

Ke Yang

1 Introduction

Drawing on examples from Beijing, this chapter discusses the implications of mobile phone use among migrant workers in China. China has experienced rapid growth in the number of mobile phone users during the past decade. Up to the first of May 2007, the number of mobile phone users reached 487 million.[1] As the cost of both mobile phones and phone usage fees have steadily come down,[2] more and more people from lower social and economic strata have begun to use mobile phones. These developments provide the background for this research into mobile phone use among migrant workers.

This chapter is based on a survey of mobile phone usage in Beijing carried out during the autumn of 2006.[3] In the course of the survey, our team interviewed 487 people about patterns of mobile phone use. Among these interviewees, 178 were migrant workers. 71.8% of the 178 migrant worker interviewees were male, 87.4% were under 30 years old, 64.6% had never been married, and 59.7% of them earned less than 1,000 *yuan* per month. These migrant workers were typically employed as construction workers, security guards, vehicle maintenance workers, maids,

[1] Data source: http://www.iimedia.cn/3/35/200766/093513.html

[2] Taking as an example a widely used China Mobile (Beijing) service plan, the fee for making a local call can be as low as 0.12 *yuan* per minute, while the fee for receiving calls can be almost free if a monthly fee of 6 *yuan* is paid. Data source: http://m-zone.bj.chinamobile.com/mzone/ywjs/tcyw/dtjh/

[3] This research project is supported by the Department of Applied Social Sciences, The Hong Kong Polytechnic University. Besides the author and the research team members are mainly from the Department of Sociology, Peking University, China.

K. Yang (✉)
Institute of Sociology, Chinese Academy of Social Sciences, Beijing, China
e-mail: yangke@cass.org.cn

P.-l. Law (ed.), *New Connectivities in China: Virtual, Actual and Local Interactions*,
DOI 10.1007/978-94-007-3910-9_6, © Springer Science+Business Media B.V. 2012

Table 6.1 Interviewees' details

Name of migrant workers	Sex	Profession
Da Lee	Male	Remodeling worker
Xiao Lee	Male	Remodeling worker
Ah Guan	Male	Remodeling worker
Da Zhou	Male	Security guard
Xiao Zhou	Male	Security guard
Xiao Hu	Male	Security guard
Xiao Yang	Female	Lottery ticket saleswoman
Xiao He	Female	Hairdresser

hairdresser, restaurant workers, and other workers in remodeling and other similar semiskilled and unskilled service occupations. Our initial round of interviews using a structured questionnaire revealed that all 178 interviewees had used mobile phones, and 158 of them owned their own mobile phones. Subsequent to the initial questionnaire-based interviews, we conducted in-depth interviews with eight of the survey subjects. Both the initial and in-depth interviews were conducted in the interviewees' workplaces, so that we could better understand their working conditions. We believe that if we combine these in-depth interviews with quantitative data, we will be able to present a more comprehensive and in-depth account of the role of mobile phone use in the social life of these migrant workers (Table 6.1).

The discussion below is based on both the eight in-depth interviews and the written answers to three open-ended questions on the initial questionnaire. (These questions were: "In the future, what new mobile phone functions do you wish to have?" "What does the mobile phone mean to you?" and "How do you feel about working away from your hometown?") Through their stories from daily life, this chapter attempts to explicate migrant workers' mobile phone usage against a background of their alienated situation in Beijing. The first section of the chapter provides background on migrant workers' working in Beijing city. Next, in the second section, I explore the responses of the migrants toward their situation in relation to the usage of mobile phone. Specifically, I introduce and discuss four concepts for understanding patterns of migrant mobile phone use: *feigned presence*, *concern in absence*, *jianghu relations*, and *romantic relations*. In the conclusion, I go on to raise remaining problems implied in this study and discuss how they might relate to further research.

2 Alienation in Real Life

Beijing, the biggest city in northern China, is dotted with many modern buildings, where the socioeconomic and political elite dwell. One community where we conducted research is an elite residential complex housing managers and professionals associated with world-renowned multinational companies and is within walking distance of the biggest shopping mall in Asia. But for our interviewees who live in shabby basements or shanties, life is much less comfortable. Unlike the migrant workers in the factories of the Pearl or the Yangtze River Delta regions, the migrant workers we interviewed work mainly in the service industry. Thus, their lifestyle is different from the factory workers in coastal provinces.[4] As most of the migrant workers in Beijing working in the service industry do not have labor contracts, it is very common for them to work overtime, have limited freedom, and have working and life spaces which overlap. The dwellings and the lives of the migrant workers we studied fill the crevices in the metropolis' modern veneer. They exist embedded in the fractures of more mainstream communities, half hidden among the high-rise apartments and luxury shopping malls.

Most of the migrant workers interviewed work in remodeling or as security guards at luxury residential developments. In both these occupations, the migrants are in close contact with wealthy Beijing residents and witness the lifestyles of them. Migrants report that their job situations make them deeply aware of the unbridgeable gap in quality of life that separates them from the wealthy that benefit from their labor.

The alienation of the migrant experience can be clearly seen in Da Lee's story. Da Lee and his younger brother have been doing remodeling work for Beijing residents for almost 10 years. When we visited them at their work site, they were renovating a villa bought by a Beijing resident as a second home. This villa was located in a beautifully landscaped suburb of the city with ample fresh air, far from the pollution and overcrowding of downtown. The Lee brothers' job was to remodel the villa in an elegant classical Chinese style befitting its owner's wealth and position. They painted the detailed wooden latticework with care and labored with heavy stones and sand to create an idyllic garden pond, but they can never dream of actually living in a luxurious villa that their own labor created. In other words, no matter how hard they work, their work is external to them. Old camp beds and dirty bedrolls are their only belongings. When asked to talk about their feelings about working far away from home, the younger brother did not conceal his bitterness: "I'm afraid I will burst into tears…. I've experienced all the joys and sorrows of life. It's too hard to work as a migrant." His working partner, Ah Guan corrected him: "He said he's tasted all the joys and sorrows, but for us, there's no joy, only sorrow!" When talking about sorrow, Ah Guan expressed his predicament: "You can say we're suffering, but, after all, we're making money. Then again, if you say we aren't suffering, just take a look at how we are dressed!" After this, Ah Guan fell silent and gazed at the dirty bedrolls in the corner of the tiny room where they slept.

Besides the alienating nature of their work, the migrants' sense of misery also comes from working overtime and work-related limits on their personal freedom. The security guards at luxury developments are poor sojourners in the city, but they are expected to work long hours, day and night, protecting the property of wealthy Beijing residents. In one such development where we conducted interviews, two security guards in charge of the Southgate took turns standing guard during a day shift that lasted from 6:00 a.m. to 10:00 p.m. During this 16-h shift, the two guards

[4] Thanks to Mr. Liu Xiaojing for his typology of modern migrant workers.

would alternate between rest and work every 2 h. That is to say, their daytime was cut into 2-h segments. This work schedule made it almost impossible for them to involve in any normal social activities. These same security guards were also subject to round-the-clock supervision and control by their employer because they did not have the means to live away from their work site. They were subject to strict regulation of their personal behavior when both on and off duty. To be specific, they were forbidden to smoke, to gamble, to play cards, and to watch TV. And surprisingly, they are not even allowed to date. If they did not comply with these rules, their salaries would be docked, in some cases by as much as 300 *yuan* or half their monthly earnings.[5] The most uncomplaining of these security guards, Da Zhou, told us, "They control you all the time. You can't do anything! You make a little bit of money, but what good is it!?" Although Da Zhou is a thrifty person and can save some money even with his meager salary, the savings cannot make up for the feelings of powerlessness and injustice created by his working conditions.

Working under these conditions, migrant workers found that they lost not only their freedom but also their dignity. An 18-year-old security guard told us: "Just after I started working here, a women living in Building 6 said she was afraid that her car would be stolen if she couldn't find a parking place. … I told her to park her car near the South Gate where I worked night shift. She said 'no, that won't work, if my car were stolen, even if you sold yourself, you wouldn't be able to pay me back.' That made me so angry! Damn it! But what could I do? After all, she's rich!"

The migrant workers we interviewed hardly find any sense of belonging in the communities where they work. What they feel in their labor is often weariness and self-depreciation. These sentiments were apparent in the following responses reported on the questionnaires:

> I feel bad indeed. When I was a student at my hometown, I was so carefree. But now, working as security guard, the biggest problem is lacking enough time to sleep. What's more I feel discriminated against; rich urbanites look down on poorly educated migrants [like me]. I'm very angry at this. This job is meaningless (caseA149).

> Currently I work as a security guard at a residential development. The job is not very tiring but it's really boring. I have nothing to do, and it is always the same day in and day out (caseA016).

> I feel so tired and incapable of doing anything (caseA112).

> I feel humiliated. When I am walking on the street I think other people look down on us migrants (case A098).

One response even quoted a famous Song Dynasty poem by Su Dongpo:

> Ten years of separation, life or death unknown…. and no one even listens to my lonely tale.

Ironically, the workers in contemporary socialist China seem to have the same situation and share the similar feelings of what Marx described as alienation.

---

[5] According to the Beijing Statistics Bureau, in 2005, the average monthly salary of a Beijing resident was 2,734 *yuan*. However, save for the head security guard, none of the security guards interviewed earned more than 800 *yuan* per month.

For them, "work is external to the worker, that it is not part of his nature; and that, consequently, he does not fulfill himself in his work but denies himself, has a feeling of misery rather than well-being, does not develop freely his mental and physical energies but is physically exhausted and mentally debased" (Marx 1964: 124–5).

3 Migrant Worker's Responses and Mobile Phone Use

"Why am I leaving home and leading a difficult life far away?" is a question common to many of the migrant workers we interviewed.[6] Actually, as we can observe from the questionnaires and interviews, migrant workers are often perplexed as to how to evaluate their displaced lives. Along with all the pain and sorrow they experience in cities, many interviewed workers also reported seeing migration as a process of self-improvement. The following quotation is a typical comment reflecting this ambiguous evaluation: "I feel unhappy and limited. It's not so happy as life at home, but I can make friends and learn many things. For example, I've learned a lot about how to use mobile phones; I can tell different brands of autos; and I have learned more about how the world works" (case A141). Living in a metropolis, migrants are exposed to contemporary urban ways of life, technology, and patterns of consumption. New terms such as "video chat," "instant messenger," and "CDMA" pepper their talk, and thus demonstrate their deep interest in and knowledge of mobile phones and other high-tech products.

While from nineteenth-century Europe to contemporary socialist China alienation may be a constant companion of labor, technological change has bought changes to daily life. As pointed out above, mobile phones in China have, during the past few years, become sufficiently affordable that the overwhelming majority of migrants surveyed use them. Below I explore several social uses of mobile phones among migrant workers in contemporary China, and discuss how this relates to their state of alienation and how it contributes to the changes of their way of life.

3.1 *Feigned Presence*

As with other social groups, mobile phone offer migrants a tool that enhances their ability to control certain aspects of their daily lives. While mobile phone usage cannot change migrants' overall disadvantageous socioeconomic position, mobile phones are often a tool for alleviating boredom and evading and mitigating the intrusive and overbearing demands of work supervisors.

---

[6] Different migrant workers offered different types of answers. Following Wang Chunguang's view, we consider migrant workers as a differentiated group (Wang 2001), and each type of them has its own concern and characteristics. Therefore, mobile phone use may have different meanings to different types of migrant workers.

Since their working hours are often incredibly long, many migrant workers use their mobile phones to alleviate work-related boredom. A female storekeeper's favorite game is to send jokes in text messages to friends while she is watching the store. Security guard Xiao Zhou likes to chat with friends by text message when he is working the night shift. After all, mobile phones as entertainment tools are small and easier to hide than TV sets or computers. When migrant workers cannot easily change their working environment, with the help of this little toy, endlessly boring work becomes less unbearable.

The games played by one security guard we interviewed were more daring. When Xiao Hu is on the night shift, he often skips out to play cards. While Xiao Hu's coworker who is out on patrol always covers for Xiao Hu and calls Xiao Hu's mobile in time for Xiao Hu to sneak back to work without being caught by their boss. Xiao Hu and his coworkers are also required to go running every night as part of their training. Running had been unpopular with Xiao Hu and his coworkers until one night when they discovered a local disco while out for a jog. Now, each night, Xiao Hu and his coworkers simply run to the disco and enjoy their leisure time until they are alerted by a mobile phone call from a lookout that it is time to get back. This creative use of mobile phones has made the nightly run popular with Xiao Hu and his coworkers.

Through the use of mobile phones, migrants enhance their ability to control how their time is spent. Although neither Xiao Hu nor any other migrant worker we interviewed put it in these terms, some aspects of their behavior seem to be passive resistance, "tak[ing] the form of passive noncompliance, subtle sabotage, evasion, and deception" (Scott 1985: 31). Through their inattentive and feigned presence at work, migrants express their attitude toward job demands, rules, and restrictions, and thus, in a sense, they win back psychologically.

3.2 Concern in Absence

In Chinese tradition, economic cooperation and reproduction are the primary functions of the family (Fei 1998: 122–23). However, with more and more peasants coming to the cities for jobs, the functions of family are changing. Rural families that had maintained traditional views are now, through exposure via labor migration, adopting urban family norms. Instead of reproduction, emotional satisfaction is coming to play a more important role as a goal for family life (Yang 2006).

For older migrants who already have families and children in their original rural homes, migration is temporary, and their ties to rural places of origin remain strong. For these migrants, the goal of migration is clear: earn money to support the family, raise children, and take care of aging parents. These clear goals give meaning to a transient and hardship-filled existence. To maximize economic gain, many of these migrants, especially the men, chose to leave their children and other dependents at home when they seek work in urban areas. But this choice is difficult because it means bearing the pain of separation from family. Because of their "capacity to retain primary social relationships over distance, the use of [mobile] phones can well go along with … remaining tightly connected to the loved ones at home"(Geser 2004). A woman who works as a storekeeper told us: "expensive though it may be, I make mobile phone calls to my mom and dad every week. I simply want to talk to them and hear their voices." For Da Lee, the cost of making long-distance calls does not matter at all. "Though I'm not some young kid, sometimes it's hard for me to fall asleep when I think about going home. … if I don't call my family once a week I feel like there is something wrong. I always love to hear my boy call me 'dad' via mobile phone. I can't hear that here, and hearing it really makes a difference."[7] When Da Lee had only one mobile phone, he gave it to his family at home. For him, expressing his concern for his family was more important than missing potential work opportunities. It seems mobile phones help mollify a role conflict that this group of migrants experience as a result of living divided lives. This kind of "concern in absence" makes distant motherhood or fatherhood possible and helps to alleviate the pain this type of migrants experience as a result of separation from home and family.

For the younger generation of migrants who go to cities for the first time, maintaining close ties with family back home is not such an urgent concern. However, their parents back in their hometowns are often more concerned with maintaining these relations and expressing "concern in absence" for their migrant children. The need to maintain communication is often the primary reason young migrants have for purchasing their first mobile phone. Interestingly, for young migrants, the convenience of communication with parents via mobile phone can sometimes become an excuse for avoiding contact with them in person. A young girl who sells lottery tickets told us: "It's enough to make phone calls. I don't want to be scolded when meeting my parents in person."

In summary, irrespective of whether the migrants feel deeply attached to family members or not, this new technology provides a method for reinforcing and maintaining family times. For older migrants and family member at home, mobile phone-enabled communication can sooth the pain of absence, while for younger migrants, it can allow them to develop a greater degree of freedom from parental control without having to explicitly challenge the importance of parental authority.

3.3 *Mobile Phones and Jianghu Relations*

Mobile phones also help workers to develop and maintain *jianghu* relations. Jianghu is the name for the subculture counterpart to the mainstream Confucian tradition. In Chinese, *jianghu* literally means "rivers and lakes," thus connoting a wider world

[7] Questionnaire data on mobile usage shows that Da Lee contacts his work partners most often (several times per day) and only contacts his family once a week. Simple frequency can be easily quantified, but such data alone can give an incomplete or even flawed picture. The strong emotional involvement that Da Lee attached to his weekly calls home can only be appreciated when qualitative data collection methods are also employed. We might have overlooked this strong emotional function of mobile phone–based communication had we not done the second round of interview.

outside of one's immediate circumstances. The concept of *jianghu* subculture is associated with the lower classes and itinerant wandering away from ones original home.[8] The concept of *jianghu* subculture has a long history in Chinese thinking, appearing in the classic *Han Fei Zi* over 2,000 years ago.[9] While the concept of *jianghu* has influenced mainstream Chinese society, it has been especially important for migrants forced to leave their hometowns and make a living in the cities (Cai 1989; Chin 1993). *Jianghu* subculture's emphasis is on personal loyalty and valor in contrast to the main Confucian value of filial piety. In contrast to the Confucian emphasis on the primacy of kinship relations, *jianghu* subculture stresses the importance of friendship. This is reflected in the saying "Depend on your parents at home, on your friend when away from home."

For contemporary migrant workers, *jianghu* friendship is a kind of practical network. These weak-tie friendships are created and maintained through leisure time activities and are not necessarily dependent on kinship or native-place ties. Our observations indicate that many young migrants maintain *jianghu* friendships. Da Lee described having a circle of such friends, and Xiao Hu, the security guard, talked about his "drinking buddies." The loneliness and insecurity of migrant life seems to push migrants to make friends with as wide a circle as possible. Besides emotional support, such *jianghu* friendships have very practical benefits. Networks of *jianghu* friends can provide protection in group scuffles and help with finding work opportunities. In all the aspects of *jianghu* interaction, mobile phone communication can play a very important role. Mobile phones help migrants to expand and maintain such *jianghu* social ties. Xiao Hu related to us how, after being fired, he was able to get a new job in a matter of minutes. One of his drinking buddies made a call to another *jianghu* friend who was able to immediately get Xiao Hu a new job. Mobile phones and *jianghu* friends also play central roles providing protection in group scuffles or conflicts. Xiao Hu related that conflicts often occurred when he or his friends frequented the local disco. In these situations, help was only a mobile phone call away:

> If I get in trouble I just make a call on my mobile and my friends come to help me out right away. I call up and ask if their boss is out. If the boss is not around, a whole vanload of them will drive over right away.

In these cases, timing is crucial, and not having a mobile phone handy could cause a disaster:

> I didn't bring my mobile with me that day, so I had to look for a public phone in a hurry, but it took too long to find one….The call was finally made, but it was too late. Their people arrived earlier, carrying lead pipes, … they gave us a good dressing down.

In another interview, Xiao Hu commented:

> Do you know why their people arrived earlier? They brought mobile phones with them! They really have a lot more experience than we do.

---

[8] Dr. Ko-Lin Chin used the term of *jianghu* in his book *Chinese Subculture and Criminality: non-traditional crime groups in America* to refer to the underworld, that is, ethnic Chinese organized crime, nevertheless, the term does not necessarily imply criminality.

[9] As a legalist, Han Fei Zi considered *jianghu* subculture and the martial prowess of its elite (*Xia*) a potential threat to the state.

Mobile phones are so important in group scuffles that when Xiao Hu and his *jianghu* friends decide to fight, they first assign one guy to be the "messenger"; he is then the only one to bring his mobile in case they need to call for help. This tactic also has the advantage of making it less likely that anyone else's mobile will get lost or damaged in the scuffle.

Though migrants may seek protection by forming networks of *jianghu* relations, the young male workers we interviewed are quite clear that these *jianghu* friends should be kept in a safe distance. When one of Xiao Hu's friends brought a "real bad guy" over to his residence, Xiao Hu got very angry: "We all asked him 'how could you bring a thug like that over to our place? If something bad happened in the future, that guy could find you! What will you do then?" As we can see from these examples, through the use of mobile phones, migrant workers can, to some degree, selectively set up protection network for themselves. In securing and maintaining *jianghu* relations, mobile phones offer migrant workers the possibility of having a "selective presence." A mobile can be used to keep connected with *jianghu* friends, while at the same time maintaining a safe distance in these relationships. In this sense, mobile phones can be considered as a tool for self-empowerment among migrant workers.

3.4 *Increased Choices and Decreased Stability in Romantic Relations*

Migrants' use of mobile phones to develop and maintain personal relationships is not limited to the *jianghu* type of relationships described above. The convenience and privacy of communication by mobile phone gives young migrant workers many more options for developing romantic relationships than were available to previous generations. During the last 30 years, attitudes toward romantic and sexual relations in China have undergone dramatic change. Now, there is a far wider range of opinions and views on what constitutes legitimate sexual and romantic behavior. Cohabitation among migrant workers has become an increasingly common phenomenon and appears to be more accepted among young migrants than among the same age cohort in rural areas. Surveys revealed that in the coastal regions, young migrants' attitudes toward love and sex are coming to resemble those of urbanites than those of their contemporaries in their inland rural hometowns (Pan 2004).

Mobile phones allow migrants to keep in contact with a range of "friends" and from that pool select suitable partners for romantic relations. This is especially true for those migrants whose work duties leave them with little free time and for those who lack access to or the ability to use the Internet to meet potential partners. The low cost, especially of text messages, and privacy provided by mobile phones makes it much easier for migrants to maintain or develop romantic relations without drawing the attentions of others. Though Xiao Hu has a steady girlfriend who pays his mobile charges when he is out of money, he still asks for the mobile numbers of other girls he met at the disco. On Valentine's Day of this year, Xiao Hu gave his girlfriend a bunch of flowers, but then went back home to talk on his mobile for an hour with some other girls he met via the Internet. Though Xiao Hu expressed that the right to choose romantic partners should be limited to the unmarried, his married coworker would also pursue girls he met at the disco. And Xiao He, an attractive migrant female hairdresser, was happy to accept a mobile phone as a present from a male friend and then use it to flirt with other men via text messages.

The prevalence of mobile phone use among migrants enlarges their scope of choice for romantic partners, but at the same time, the increased choice and greater privacy that mobile phone use brings may compromise the stability of these romantic relationships. Xiao Hu acknowledged that his girlfriend would most likely break up with him if she found out about his flirting with and calls to other girls. "Let her go. I don't want to get married right now anyway." No matter they are aware of it or not, migrant workers are changing from traditional to modern, and during this process, mobile phone has contributed to the change.

4 Conclusion

All over rural China, migrant workers have left their hometowns and flooded into cities seeking to improve their social and economic status. This migration process exposes them to both opportunity and potential exploitation in unfamiliar urban social and physical environments. Under these circumstances, patterns of mobile phone use play a role in migrant responses to alienation, discrimination, and restrictions on their freedom of action. Migrants use mobiles to achieve a "feigned presence" and avoid the boredom and restrictions of work; they use mobiles to keep connected to the rural world and show their "concern in absence," mitigating conflict between economic and emotional necessities; through the use of mobile phones, they can develop networks of *jianghu* friends for protection and increase the scope of their romantic choices. In this sense, mobile phone usage has practical value for migrant workers. This situation differs from migrants' use of mobile phones for conspicuous consumption and display reported by Law and Peng (2006) in Guangdong Province. Most of the migrants we interviewed did not use expensive mobile phones and used mobiles mainly as individual communication tools and not for display or to achieve status. It appears that mobile phone use among migrants provides this often exploited and controlled population with the freedom and convenience to communicate with any one, in any place, any time. In this sense, mobile phones play a liberating role in the lives of migrant workers. However, it is not entirely clear that this new apparent freedom is all that it seems to be. How will this expanded communicative freedom affect young migrants? Might this individual communicative tool only intensify the inclination toward self-centeredness? And if that is the case, might the abuse of mobile communication turn into another source of alienation? These issues provide potential avenues for further research and discussion.

References

Cai, S. (1989). *Zhongguo mimi shehui* [Secret societies in China]. Hangzhou, China: Zhejiang Renmin Chubanshe.

Chin, K. (1993). *Banghui yu huaren ciwenhua* [Chinese subculture and criminality: Non-traditional crime groups in America]. Taibei, China: Taiwan Shangwu yinshuguan Gufen youxian gongsi.

Fei, X. (1998). *Xiangtu zhongguo shengyu zhidu* [Rural China and child-bearing institution]. Beijing, China: Peking University Press.

Geser, H. (2004). Towards a sociological theory of the mobile phone. From http://socio.ch/mobile/t_geser1.htm. Accessed 29 Apr 2011.

Law, P., & Peng, Y. (2006). The use of mobile phones among migrant workers in southern China. In P. Law, L. Fortunaiti, & S. Yang (Eds.), *New technology in global societies* (pp. 245–258). Singapore: World Scientific.

Marx, K. (1964). *Early writings* (T. B. Bottomore, Ed. & Trans.). New York: McGraw-Hill.

Pan, S. (2004). *Dangdai zhongguoren de xingxingwei yu xingguanxi* [Sexual behavior and relation in contemporary China]. Beijing, China: Sheke Wenxian Chubanshe.

Scott, J. C. (1985). *Weapons of the weak: Everyday forms of peasant resistant*. New Haven: Yale University Press.

Wang, C. (2001). The relationship of new generation of rural floating population's social identity and merger of rural and urban. *Shehuixue Yanjiu, 16*(3), 63–76.

Yang, S. (2006). *Jiating shehuixue* [Sociology of family]. Beijing, China: Gaojiao Chubanshe.

Part III
The ICTs and Migrant Workers in Southern China

Chapter 7
Mobile Cultures of Migrant Workers in Southern China: Informal Literacies in the Negotiation of (New) Social Relations of the New Working Women

Angel Mei-yi Lin and Avin H.M. Tong

1 Introduction

Since the mid-1980s, many transnational corporations have rushed into mainland China for both its promise of large markets and low labor and production costs. Many rural Chinese people have also dreamed of living a city life through working in the new factories opened up by transnational corporations in urban areas, and in large numbers they have been moving from villages to urban cities, starting massive flows of rural-to-urban migration. These demographic flows have initiated chains of radical sociocultural changes in China in the past two decades (Gaetano and Jacka 2004; Lin 1997; Ma and Cheng 2005; Solinger 1999). Many scholars have conducted research on migrant workers to study their working environments, social networks, and consumption practices. In this section, we will review some of the major studies to provide a general picture of the social lives of migrant workers in modern-day China.

1.1 Rural-to-Urban Migration: An Individuating Project Driven by the Desire for an Urban Middle-Class Life

A new class of workers has arisen as China is evolving into an industrial powerhouse, and these workers desire to transform their rural lives by selling their hard labor in the newly formed industrial markets. In particular, rural women started to

A.M.-y. Lin (✉)
Faculty of Education, The University of Hong Kong,
Hong Kong, China
e-mail: angellin@hku.hk

A.H.M. Tong
Faculty of Education, The Chinese University of Hong Kong,
Hong Kong, China

P.-l. Law (ed.), *New Connectivities in China: Virtual, Actual and Local Interactions*,
DOI 10.1007/978-94-007-3910-9_7,

become important labor in factories and service sectors. Pun (2005) studies a group of women workers in China and finds that they are proud of their newly found urban identities. Through working in the factory, they seem to feel that they are at least active agents in living their own lives, and their wage-earning power seems to give them some sort of new power and freedom (Pun 2005). In Ma's study (2006), those factory women are happy to get away from hometowns, so that they can marry later and develop their own career. This creates a desire for women to leave the village and escape to the urban city, to reconfigure both women's identities and Chinese familial relations.

To these women, the first step to realize one's dream to attain a modern life style is to enter the factory, although this also means immersing oneself in hard and tedious production work. Pun (2005: 9) describes the process of entering the factory at the beginning as actually a process of "individuating the self," which makes someone realize that "it had recourse to nobody but itself." Although the work is simple, it is repetitive and monotonous, and everybody lives up to a strict schedule without much support from one another (Ma 2006; Pun 2005). Ma (2006) finds that workers cannot refuse to work overtime, and they need to comply with the machine's rhythm, and not with one's own biological clock. And consumption, for these workers, seems to be a way to gain temporary relief and to escape into fantasies about modern lifestyles and identities. We shall discuss more about this in the next section.

1.2 Popularity of Mobile Phones and Text Messaging: Markers of Membership in Modern Communities

The mobile cultures developed in China can be seen as one of the ways of acquiring a modern lifestyle through consumption. Law and Peng (2004a, b, 2007) report on the widespread use of mobile phone and SMS among migrant workers in southern China. According to statistics released by the Ministry of Information Industry,[1] the number of mobile phone subscribers reached 305 million in mid-2004. The national average penetration rate was 24%, with the Guangdong Province having an above-national average penetration rate of around 60%. A recent survey of Law[2] indicates that 64% of the migrant workers reported the use of mobile phones, and over 40% of them reported using it for text messaging.

Migrant workers are among the lowest-income groups in Guangdong, largely taking up skilled and semiskilled jobs in factories with monthly salaries ranging from 400 to 800 *yuan* (US $1 is approximately 8.23 *yuan*) (Law and Peng 2004a). However, a migrant worker is willing to spend double or triple her/his monthly

---

[1] See http://www.mii.gov.cn.

[2] An initial figure obtained from an unpublished project of Law on the situation of migrant workers in China

salary to purchase a mobile phone. The mobile phone seems to have become a symbol of social identity and a marker of the keenly desired urban, cosmopolitan lifestyle. In particular, SMS has become very popular in China. In 2001, 18.9 billion messages were sent in Guangdong; in 2002, the number of messages increased strikingly to 90 billion, averaging 247 million a day.[3] In our study, most of the workers reported sending/receiving around 100–200 messages per month.

In the research literature, SMS has been reported to be an ideal tool for making social arrangements and recording appointments and addresses (Haig 2002). Cartier et al. (2005: 9) use the term "information have-less" to refer to a "social, economic, and political category for millions of rural-to-urban migrants and laid-off workers who populate the vast gray zone of China's digital divide." Those low-end digital technologies including SMS play an important role in enhancing labor mobility (both physical and social) and in the formation of "translocal networks." These technologies can perform critical informational functions, as tools of communication that keep migrants connected with family members, hometowns, and also new friends and contacts, which enables and facilitates information flows (Cartier et al. 2005). Law and Peng (2004a, b, 2007) describe a number of both social and practical functions served by mobile phones and SMS, for example, maintaining contact with family members in their home village, maintaining contact and friendship with kinsmen/women, and for purposes of dating and courtship.

After outlining the recent research literature above, in the next section, we shall introduce our interview study of a group of migrant workers in southern China. Then important findings will be presented and their implications will be discussed. In the last part of the chapter, directions for future studies will be explored.

2 The Present Study

The present study examines mobile cultures among migrant workers in the industrialized villages of Tangxia and Humen Towns of Dongguan in the Guangdong Province in southern China. A total of 63 migrant workers, including 38 males and 25 females (pseudonyms are used to protect the privacy of informants), were interviewed between July 2004 and March 2005. The informants were young villagers from less-developed or poor provinces such as Hunan, Sichuan, and Guangxi. Tangxia is one of the 32 towns under Dongguan's administration. It has 20 villages, with around 33,000 local residents and over 600,000 workers from outside the town. The factories there produce various products such as plastics, toys, and garments.

Most of the informants have finished their junior middle schooling and have a monthly salary between 400 and 800 *yuan*. Some of them have spent 7 or more years in Dongguan, while some have just been there for a year. In-depth semistructured individual and group interviews with the migrant workers were conducted to

[3] "Thumb Economy" Earned 9 Billion, *Mingpao Daily* (2003, September 14), B6.

analyze their mobile consumption practices, as well as to understand their work life, leisure life, and dilemmas/conflicts that they might experience in their everyday social relations. Most interviews were conducted in factories, and some interviews were conducted in restaurants when the interviewers had dinner with migrant workers.

While both male and female workers were interviewed, our analysis in this chapter will focus on understanding the role played by SMS in facilitating the female workers' development of informal literacies, which in turn facilitates their negotiation of new social and gender relations.

3 Findings

We shall divide our findings into two main sections. In the first section, we discuss how these workers, limited as they seem to be in their formal literacy training, mobilize their linguistic resources with the help of SMS manuals (Lin 2005) to become creative SMS writers. Then in Sect. 3.2, we shall discuss the role that SMS plays in helping female workers to negotiate new kinds of virtual social relations.

3.1 The Creative SMS Literacies of Migrant Workers: Mobilizing Limited Linguistic Resources to Craft Out Arty Messages

3.1.1 Emerging SMS Literacies Among Workers with Limited Formal Literacy Skills

The existing statistics show that 18.9 billion messages were sent in 2001 in Guangdong, but in 2002, the number of messages increased strikingly to 90 billion, averaging 247 million a day.[4] In our study, most of the informants have come from villages and had little formal education (except for a very few of them, who have received college education). In doing their messaging, the migrant workers seem to need some language support. They report readily turning to SMS manuals (see Fig. 7.1), which are published as handy, smaller-than-pocket-size booklets selling cheaply (e.g., 3 *yuan*) and circulating widely among these migrant workers (Lin 2005). These manuals are organized as lists of numbered entries of short message texts placed under different topic headings (e.g., "humorous sex jokes," "true sentiment confessions"). It seems to offer a textual supermarket of diverse linguistic and discursive resources as well as gender ideologies (Lin 2005) made available to the migrant workers when they use SMS for purposes which are related to seeking (virtual) relationships and (virtual) experience of romantic/sexual pleasure and excitement.

[4] "Thumb Economy" Earned 9 Billion, *Mingpao Daily* (September 14, 2003), B6.

Fig. 7.1 SMS manuals popular among migrant workers in southern China

Some workers report making efforts in creating their own messages, and they sometimes get inspiration from their friends' messages and also SMS manuals. For instance, Mr. Chow admits that he will copy some words/sentences from the SMS manuals, especially for romance topics. On the other hand, the boyfriend of Miss Fung usually copies passages from the newspaper (e.g., feature articles), which are often related to romance and jokes. Research on intertextuality shows that literacy artifacts and images invoke social practices from a range of different social contexts. In our study, the messages provided in SMS manuals seem to serve as important sources of references to SMS users in their everyday texting practices.

Although SMS manuals are quite popular among migrant workers, some of our informants claim that they do not rely on manuals and prefer creating messages themselves. Miss Fung, for instance, criticizes those "typical examples" in SMS manuals as "not interesting." Another informant, Miss Mui, insists that she just reads SMS manuals but never copies messages from them, "I seldom send out those messages, (what I send out) are all my own words." One of the male workers, Mr. Yu, says, "Those (messages) written on the manuals are already outdated. Those messages that come out from one's mind are the best...think and create a message; it's a kind of training."

Miss Mui has bought a few SMS manuals, which teach people how to write different types of messages such as jokes and interesting stories. But she insists that she uses the guide very occasionally. She describes some tactics in "recreating" messages received/learned from the SMS manuals: "I seldom copy every word of it; I just take it as a reference…change it to make it better…won't forward it without changing a word." She thought that the "recreation" process is rather easy, as she puts it:

> It's easy for me to send (a message), and it's easy for me to type, very fast. Even when others sent me a copied message, I will change it a bit when I send it to others; add some more or delete some parts…just add some new words in the front or at the back, or take away some of it.

Researchers in informal literacy believe that the schooling tradition of cultivating formal literacy seems to downplay the role played by the indigenous, informal linguistic and creative resources of people (Street 1993). In our study, it is interesting to find that workers with relatively limited formal education sometimes display great creativity when they compose mobile messages. For instance, a male worker, Mr. Ho, says, "Some of my friends enjoy writing poems, and sometimes (they) write good wishes and send (them) to me." Writing a poem is not an easy task for a factory worker, as most of them only have limited schooling experiences. In fact, factory work does not involve literacy work, and there is no chance for migrant workers to develop their literacy skills. SMS thus provides them with a valuable medium and opportunity for practising their informal, everyday literacies.

Literacy learning and uses are actually integrated into everyday activities (Hamilton 2006). Ethnographic studies of everyday settings (Barton and Hamilton 1998) have explored the relationship between the lived experience of literacy and learning. One of the important notions of everyday literacy research is to see literacy as a social practice (e.g., choosing verses in greeting cards exchanged, cherishing letters and autographs). We thus need to look into the diverse ways in which literacy is acquired, not just in school settings and by children but also in everyday life and by adults. In our study, it seems that through creating and recreating different types of SMS, it is possible for migrant workers to acquire and practise literacy skills informally in their everyday life.

3.1.2 SMS Literacies in the Negotiation of (New) Gender Relations

Men and women seem to express themselves differently, as revealed in their practices of using mobile phones/SMS. As reported in Law and Peng (2004b, 2007), many male workers claim that they would prefer making voice calls to sending messages, while female workers prefer sending messages. This point is also made by our female informant, Miss Fung, as her boyfriend seems to have no interest in sending SMS. She recalls some unhappy experiences, "If I send him messages, ask him what he is doing, he will reply: 'I am busy'…every time he says he is busy after chatting for a few sentences…Every time it's he who uses 'busy' as a reason for closing the conversation." She also prefers more expressive writing styles instead of instrumental and practical ones. She says, "He never seriously sends me three words saying 'I miss you,' he never sends it…I know he misses me…but he never sends (me) those three words."

However, some male informants are actually willing and eager to use this new medium of communication to express themselves and to find themselves a (virtual) social life. A male informant, Mr. Ho, admits that SMS has become an important part of his leisure life, and he will feel depressed if no one sends him any SMS. "If I haven't received any messages for a few days, I will feel a bit unhappy in my heart. (I) will have a feeling of being isolated." A similar remark is made by Mr. Au, "If I haven't received any SMS for two or three days, I will ask myself: how come I haven't received any SMS?" Besides, some male workers have become heavy users of SMS. A female worker, Miss Lau, recalls, "I know one boy who is fond of sending SMS; each time (the message) he sent me contains two hundred to three hundred words." These new attitudes and behavior of some male workers seem to deviate from the stereotypical image of Chinese men, who are often thought to be few in words especially when it comes to social writing.

In some cases, male workers make use of SMS in settling arguments with girlfriends, and some of them, like Mr. Kok, use SMS for dating and courtship in the initial stage, as they report being able to express their feelings more freely in a textual world. As Zubair (2001) suggests, we should look at literacy practices as part of the broader gender relations situated in the larger social and cultural contexts. In our case, informants reveal both confinement to certain gender stereotypes (e.g., writing styles) and some breakthroughs from some traditional expectations (e.g., efforts of some males in creating messages, and their high frequency of sending long messages). Our preliminary interview data thus point to the possibility that some male workers are developing and using their SMS literacy skills to negotiate and work on achieving more satisfying social and gender relations. In the next section, we shall discuss, in particular, how female workers have capitalized on their newly developed informal SMS literacy skills to acquire more social capital.

3.2 *Seeking and Constructing Virtual Romance Through SMS Literacy Practices*

3.2.1 Acquiring Social Capital: Making Oneself Attractive in the Virtual Space

Since the mid-1980s, descriptions of literacy as a form of cultural capital have become an integral part of the debates over social effects (Bourdieu 1991; Cook-Gumperz 1986). Bourdieu (1986) has theorized different types of capital in different social fields. Cultural capital describes "the sum total of durable knowledges and practices, discursive and material resources acquired by individuals as they develop across their life trajectories," while social capital refers to the "direct access to social and cultural institutions and organizations". In our study, migrant workers actually enter a complexity of social fields and involve in complex interrelationships between cultural and social capital through their use of mobile phones/text messaging.

Mobile phones provide an online service for people to meet new people in the cyberspace. Workers are always receiving messages from strangers hoping to make romantic contact with others through mobile phones (Law and Peng 2007). Some migrant workers choose to chat with strangers (especially those of the opposite sex) using this online service provided by the mobile phone service. There is a popular network called "Mobile Fantasy Network" organized to help men and women find "good friends," and in most cases "lovers." To develop such kind of intimate relationship, cultural capital is required to make oneself more appealing and sociable. The most illustrative case is that of Miss Fung, who started participating in the chat room a few years ago and has now become very "professional" in searching "suitable" male virtual partners. She likes using different names, such as "Apple," "Ice," and "Windie" to log on to different chat rooms. She spends quite a lot of time on creating beautiful "names" for herself in the chat room, which involves certain literacy skills. She analyzes her naming practices in this way, "People like pretty girls. I have once been called Ching Chi ("Virtuous Girl") and also Han Bing ("Cool Ice")."

Miss Fung shows her social skills and SMS expertise by readily providing others with tips on how to create impressive and favorable identities. For example, she teaches male players (in the SMS chat room) to claim themselves as young men with a cheerful character and handsome appearance so as to attract female netizens. Here is the personal experience of Miss Fung, which points to her linguistic tactics in constructing attractive online identities:

> Men usually claim to be twenty-something, it is because if they are already in their thirties and still (need to) go online, then they must have some problems. Others don't want to chat (with you)…it will fail if you are too honest. (People) want to kill loneliness when (they) participate in online chatting, and if you are too boring, then I prefer not to chat (with you) instead…sometimes I ask others, are you handsome? When he says he is not handsome, then I will say sorry (to him), I don't want to chat with you.

Proposes the "personal growth hypothesis," which links the acquisition of literacy and series of consequences. He proposes that individual "voice" and expression will lead to psychological development. This means that literate success is "a matter of expressing individual difference, establishing self-esteem, identity and 'choice'". The case of Miss Fung illustrates the discursive tactics of building an impressive self-identity by utilizing the informal SMS literacy skills that she has developed. The beautiful names and communication tactics are highly rewarding emotionally, and they also hold a promise of meeting new friends (in particular, friends of the opposite sex). In theoretical terms, she is converting her cultural capital (i.e., informal SMS literacy skills) into social capital (e.g., social networks) in the process of text messaging.

3.2.2 Fantasies Woven Around Romantic and Ambivalent Relations

Ellwood-Clayton (2003) finds that some Filipinos are restructuring amorous lived relationships and creating virtual ones through text communication, and so text messaging seems to have induced changes in the traditional courtship practices among young people. SMS seems to provide a space for emerging amorous feelings and romance, and it is found to instill personal bravery among its users. In particular, women are encouraged to speak out their feelings and take more initiative in the textual world. Though many informants claim that they are not purposefully searching for romance, most of them admit that it is much more fun to chat with netizens of the opposite sex.

This kind of ambivalent relationship and virtual romance provide the workers with daydreams and fantasy, which can be useful in consoling their lonely spirits. Gossip, jokes, and laughter centered on love and sex helped the workers to cope with the tedious and difficult factory life, as a weapon against the alienation of work (Pun 2005). Some women adopt new literacy genres to open up new worlds and identities, to overcome their oppressive situations. In our study, some female workers try to attract strangers of the opposite sex by creating beautiful names and applying communication tactics, and in return they receive some sweet phrases that they are longing for. It is similar to the kind of "fantasy literacy" suggested by Luke and Carrington (2001: 17), which aims at "a suspension of position in the social fields and scapes of globalisation and a psychic disengagement with flows." It enables the SMS users to read and write out of local place and space, to enjoy romance with complete strangers beyond geographical boundaries.

3.2.3 Negotiation with Love and Gender Relations in Reality

This kind of pleasure obtained from virtual romance fulfills workers' (especially young women's) desires for romantic love. According to what our informants told us, some of our workers are dissatisfied with their love life in reality, for example, having communication problems with their lovers. As Miss Fung points out in the above examples, she enjoys the "sweet words and honeyed phrases" that are "absent in reality." In fact, constructing virtual romance may have an influence on real-life love relations. In the case of Miss Fung, her boyfriend dislikes her spending so much time and money on sending SMS/chatting with other men. Miss Fung recalls his words:

> He told me, I dislike you chatting on the net. I ask why, he said it particularly hurt a relationship...He believes that all I heard online are sweet words and honeyed phrases, which cannot be heard in real life...it is because things on the net are so insincere; and (they) hurt our relationship. But sometimes when I feel annoyed alone, I will want to chat with someone.

Mediated communication always underlies specific restrictions of coding. Walter (1996) suggests the term "hyperpersonal interaction," which means that mediated communication is more desirable as compared to face-to-face interaction (Hoflich 2003). This kind of "hyperpersonal interaction" enables migrant workers to develop ambivalent and even intimate "online" relations with netizens, which seem to be more pleasurable and enjoyable than the "offline" love life. Nevertheless, as indicated in the case of Miss Fung, virtual romance (originating from the online world and then extended to other communicative means, like SMS in this case) satisfies the desires for being loved in the workers' hearts but also creates communicative problems (e.g., wrong expectations, frequent quarrels) between "real lovers" in reality.

One important phenomenon we see here is the changing life aspirations, both in career and love, of the young women in China. Nowadays, many young Chinese rural women no longer stay at home and wait for a matchmaker to decide their fate; instead many of them are encouraged to go out and leave their villages and look for their own love and life (Pun 2005). In our study, it is found that women workers often have more job opportunities than men, and they even make more money than their male counterparts. Our informant, Miss Hang, points out, "Male work is more difficult (to find) than female work right from the beginning…work outside always prefers female labor." The rising social status and increasing economic power of women grant them more freedom in marriage and also in pursuing one's life goal. They are both eager and confident to choose a better partner whom they really love.

These young working women will have higher expectations on men, as their qualities have improved and their social circles have enlarged. Miss Sang, a factory owner, believes that it is difficult for a female worker to find a suitable male partner in the workplace. She says, "If you ask our (female) workers to choose a man in the factory, (they) absolutely cannot choose one. These men are poor and not good-looking either." With the experience of working in the city and exposure to popular culture, liberating messages found in SMS manuals like "developing one's own career," "finding true love," and "enjoying one's own life" have become mottos for these modern working women with new aspirations.

4 Conclusion: Alternatives for Women—The Potential of Literacy as Empowerment

In southern China, compressed and multilayered modernity means the pluralization of life choices in which various forms of individuality can be imagined and practiced (Ma and Cheng 2005). Migrant workers, who have been accustomed to a relatively stable set of practices of early marriage in the villages, are now thrown into a fluid set of discourses about dating, love, romance, choices, and desires. As Ma and Cheng (2005) point out, "In their de-traditionalized and de-territorialized lifeworlds, migrant workers find themselves in a ruptured discursive space where new intimate experiences require new hybrid vocabularies to express themselves." In our interviews, we found that the young female migrant workers need to seek fantasies and pleasures in their leisure life—ambivalent relations and virtual romance and strengthened or expanded social relations—to balance the alienating and individuating factory life, as well as to live a completely new woman's life as compared to the old times in rural areas.

In this modern, urban, alienating workplace, women have been presented as the major victims in many academic discourses. For instance, Pun (2005) is rather pessimistic in concluding about the future life of the female migrant workers, as she finds them under multiple and inescapable oppressions. However, some people believe that literacy can empower, through collective actions and enhancement of individual capacity (Rockhill 1993). We propose that literacy is highly related to power relations and can offer alternatives in people's lives. In our study, through the use of mobile phone and SMS, these migrant workers in southern China search for virtual pleasure and emotional release. In particular, women workers seem to have developed a desire to seek pleasurable leisure lifestyles outside of the workplace. Informal literacies are both acquired and strengthened in their frequent SMS communication practices, which involve different dimensions of competencies ranging from technological skills to creative writing and to communicative tactics in fashioning an attractive self on the net.

The study of Rockhill (1993) demonstrates how women talk about their longing for literacy. Robinson-Pant (2001) argues that women themselves tend to associate literacy with a new identity and want to demonstrate their new roles as "literate" women. In this study, it also seems that informal literacy practices have become essential tools in dealing with the urban city discourses and to pursue a "modern life" of one's desire even with very limited means. Our female informants have already shown their dreams/desires to escape from the patriarchal familial relations in the rural village; they dream of developing a new urban identity together with new, modern womanhood. It seems that these factory women have acquired new attitudes toward love, marriage, and members of the opposite sex. These attitudes seem to be very different from the previous values of traditional Chinese women. Informal literacy seems to offer them some cultural capital to engage in different communicative practices (e.g., composing poetic SMS, fashioning attractive poetic names for oneself), which may reward them with both a sense of agency and with social capital in finding new pleasurable social relationships, even though these might be transient, virtual ones.

Our research also leads us to a new approach to studying mobile communication. This approach sees mobile communication, text messaging in particular, as a social practice involving everyday literacies. The study of everyday literacy practices draws attention to the texts of everyday life and the texts of private and personal life, just like some of the intimate text messages found in our study. It seems that people can utilize and improve their literacy skills (e.g., technological skills, writing and reading skills) through text messaging while strengthening old relations and establishing new (virtual) connections. This reminds us to extend the scope of "literacy" beyond the school setting and to study new media literacy practices embedded in everyday social interactions.

Further studies can research on other informal literacy practices, to uncover their meanings and impact on people's lives situated in specific sociocultural contexts. The different cultural and social practices of male and female workers can also be studied, to help them explore more about their own needs and situations and to enable them to build more satisfying gender relations in the rapidly changing Chinese society. Much more work, both of the research and social activist kinds, would seem to be needed in the near future.

Acknowledgments The authors are indebted to Pui-lam Law, Jing Wang, Ke Yang, Xiaojing Liu, and Yinni Peng for their valuable work in collecting the interview data for this study. Special thanks also go to the anonymous reviewers for their very useful comments and suggestions.

References

Barton, D., & Hamilton, M. (1998). *Local literacies: Reading and writing in one community*. London: Routledge.

Bourdieu, P. (1986). The forms of capital. In J. G. Richardson (Ed.), *Handbook of theory and research for the sociology of education* (pp. 241–258). New York: Greenwood Press.

Bourdieu, P. (1991). *Language and symbolic power* (J. B. Thompson, Ed.; G. Raymond and M. Adamson, Trans.). Cambridge: Polity Press.

Cartier, C., Castells, M., & Qiu, J. L. (2005). The information have-less: Inequality, mobility, and translocal networks in Chinese cities. *Studies in Comparative International Development, 42*(2), 9–34.

Cook-Gumperz, J. (1986). Literacy and schooling: An unchanging equation? In J. Cook-Gumperz (Ed.), *The social construction of literacy* (pp. 16–44). Cambridge: Cambridge University Press.

Ellwood-Clayton, B. (2003). Virtual strangers: Young love and texting in the Filipino Archipelago of cyberspace. In K. Nyiri (Ed.), *Communications in the 21st century. Mobile democracy: Essays on society, self and politics* (pp. 225–239). Hungary: Passagen Verlag.

Gaetano, A., & Jacka, T. (Eds.). (2004). *On the move: Women in rural-to-urban migration in contemporary China*. New York: Columbia University Press.

Haig, M. (2002). *Mobile marketing: The message revolution: A cracking insight into how to really get personal with your customers*. New York: Kogan Page.

Hamilton, M. (2006). Understanding the everyday: Adult lives, literacies, and informal learning. In A. Mckeough, L. M. Philips, V. Timmons, & J. L. Lupart (Eds.), *Understanding literacy development: A global view* (pp. 137–152). Mahwah: Lawrence Erlbaum Associates.

Hoflich, J. R. (2003). Part of two frames: Mobile communication and the situational arrangement of communicative behaviour. In K. Nyiri (Ed.), *Communications in the 21st century. Mobile democracy: Essays on society, self and politics* (pp. 35–53). Hungary: Passagen Verlag.

Law, P., & Peng, Y. (2004a). Cellphone, Internet, and the SARS epidemic. *Proceedings of the International Workshop on Mobile Technologies and Health: Benefits and Risks*. Udine, Italy, 2004, 7–8.

Law, P., & Peng, Y. (2004b). The use of cellphones among migrant workers in southern China. *Proceedings of the Conference of Mobile Communication and Social Change*, 18th–19th October, Seoul, Korea.

Law, P., & Peng, Y. (2007). Cellphones and the social lives of migrant workers in Southern China. In R. Pertierra (Ed.), *The Social construction and usage of communication technologies: Asian and European experiences* (pp. 126–142). Quezon City: University of the Philippines Press.

Lin, G. (1997). *Red capitalism in southern China: Growth and development of the Pearl River Delta*. Vancouver: University of British Columbia Press.

Lin, A. (2005, June 7–8). *Romance and sexual ideologies in SMS manuals circulating among migrant workers in southern China*. Paper presented at the Mobile Communication Conference. City University of Hong Kong, Hong Kong.

Luke, A., & Carrington, V. (2001). Globalization, literacy, curriculum practice. In R. Fisher, M. Lewis, & G. Brooks (Eds.), *Language and literacy in action* (pp. 231–250). London: Routledge/Falmer.

Ma, E. (2006). *Jiu ba gong chang: Nan zhongguo cheng shi wen hua yan jiu (Chinese)*. Nanjing: Jiangsu ren min chu ban she.

Ma, E., & Cheng, H. (2005). Naked bodies: Experimenting with intimate relations among migrant workers in south China. *International Journal of Cultural Studies, 8*(3), 307–328.

Pun, N. (2005). *Made in China: Women factory workers in a global workplace*. Durham/Hong Kong: Duke University Press/Hong Kong University Press.

Robinson-Pant, A. (2001). Women's literacy and health: Can an ethnographic researcher find the links? In B. Street (Ed.), *Literacy and development: Ethnographic perspectives* (pp. 152–170). London/New York: Routledge.

Rockhill, K. (1993). Gender, language and the politics of literacy. In B. Street (Ed.), *Cross-cultural approaches to literacy* (pp. 156–175). Cambridge: Cambridge University Press.

Solinger, D. (1999). *Contesting citizenship in urban China: Peasant migrants, the state, and the logic of the market*. Berkeley: University of California Press.

Street, B. (Ed.). (1993). *Cross-cultural approaches to literacy*. Cambridge: Cambridge University Press.

Walter, J. B. (1996). Computer-mediated communication: Impersonal, interpersonal and hyperpersonal interaction. *Communication Research, 23*, 3–43.

Zubair, S. (2001). Literacies, gender and power in rural Pakistan. In B. Street (Ed.), *Literacy and development: Ethnographic perspectives* (pp. 188–204). London/New York: Routledge.

Chapter 8
Internet Use of Migrant Workers in the Pearl River Delta

Yinni Peng

1 Introduction

With the wide and rapid diffusion of the Internet into our societies, the social impacts of Internet use draw the attention of academia. Compared with the unequal access to the Internet, the differentiated use of the Internet becomes more important in the eyes of most scholars recently (Howard et al. 2001; Haythornthwaite and Wellman 2002; DiMaggio et al. 2003; Kraut et al. 2002; Zhao 2006). They state that different online activities contribute to different effects on people's social lives. Moreover, some scholars (Haythornthwaite and Wellman 2002; Katz and Rice 2002) argue that Internet use is not isolated from our real lives but grounded in people's social lives. Online activities are related to the social contexts in which Internet users are embedded. Therefore, it is important to examine the diversity of Internet use and interpret the social implications of differentiated Internet uses within the social context which Internet users experience. This chapter discusses how the different uses of the Internet among migrant workers affect their social lives in the Pearl River Delta (PRD). First, it investigates the different patterns of Internet uses among migrant workers in the PRD. Second, it explores the impact of the Internet on migrant workers' social networks. Third, it discusses the Internet's role as a release from the pressures of work among migrant workers.

2 Background and Research Method

The PRD has the largest pool of migrant workers in the world's factory—mainland China. It is the site of 33,000 corporations, which are the dominant world's suppliers of light industrial products such as consumer goods, garments, and accessories.

Y. Peng (✉)
Hong Kong Shue Yan University, Hong Kong, China
e-mail: pengyinni365@gmail.com

P.-l. Law (ed.), *New Connectivities in China: Virtual, Actual and Local Interactions*,
DOI 10.1007/978-94-007-3910-9_8, © Springer Science+Business Media B.V. 2012

A large number of young peasants from less-developed provinces, such as Sichuan, Hunan, Guangxi, and Jiangxi, have sought jobs in the PRD (Li and Tan 2000; Wang 2001; Law and Peng 2006). In 2000, there were more than 25.3 million migrant workers in Guangdong province, and the PRD accounted for 55%.[1] By January 2005, it was estimated that the number of migrant workers in Guangdong province had reached 42 million.[2] Most migrant workers in the PRD work in labor-intensive factories established by foreign and local entrepreneurs (Li and Tan 2000; Wang 2001; Law and Peng 2006). They belong to the lowest income group in the PRD. In 2004, the average monthly salary of migrant workers in the PRD was about 964.8 *yuan* (including overtime pay), which represents 57.9% of an average urban worker's salary in Guangdong province.[3]

According to a survey conducted by Peking University, in 2006, over 36% of migrant workers in the PRD went to Internet cafés at least once every 5 days, and more than 21% used the Internet everyday.[4] Their average monthly Internet expenses were about 68 *yuan*, which is equal to one-fifth of their consumption.[5] It is clear that the Internet has a strong attraction for migrant workers. It would be illuminating to explore the effects of Internet use on migrant workers in the PRD.

This chapter is based on a qualitative research conducted in Shenzhen from March to June 2007. Shenzhen is one of the eight economically advanced cities in the PRD. In 2005, migrants accounted for 6.46 million of its 8.28 million population. Shenzhen is a city of migrants. The spread of the Internet in Shenzhen has been rapid. By the end of 2006, the number of Internet subscribers in Shenzhen was reported to be over 2.15 million.[6] It is, therefore, appropriate to choose Shenzhen as the focus of an investigation of the relationship between the Internet use and the social lives of migrant workers. We conducted our field work in two Internet cafés: Xin Internet café and Jian Internet café. Twenty-six Internet users were interviewed, including 24 migrant workers (15 female and 9 male), one Internet café owner (male), and one Internet café staff member (male). Most of the migrant workers interviewed in this study come from four provinces: Hunan (6), Sichuan (4), Guangxi (5), and Hubei (5). The majority worked in factories manufacturing electronic devices, garments, or toys. Over half were shop-floor workers, and nine were either foremen or clerical staff. The average monthly salary of the participants was about 1,000 *yuan*. Their ages range from 18 to 23. They represent the typical migrant worker in the PRD. The interviews focused on their social lives and Internet use. All the interviews were semi-structured in that they were guided by a set of key questions about migrant life (family background, migration history, current work

---

[1] See National Bureau of Statistics of China (2003)

[2] Data source: http://media.163.com

[3] See Guangdong Statistic Bureau (2005)

[4] Data source: 2006, *The report of migrant workers and their mobility status in the PRD*. Department of Sociology, Peking University

[5] Ibid.

[6] Data source: http://www.nmxxb.gov.cn/nmxxb_list.asp?Unid=3393

and life state) and Internet use (consumption, web activities, social relation through Internet use). Based on the preliminary data, migrant workers were asked to provide further details about the impact of Internet use on their migrant lives in Shenzhen. The researchers also interviewed one Internet café owner and one Internet café staff member to enrich the narrative perspective and get some general information about Internet use of migrant workers in the Internet cafés. All interviews were tape-recorded and fully transcribed. In addition, field observation supplemented the research data. As most interviews were conducted in the Internet cafés, the researchers had the opportunity to observe migrant workers' Internet use. Over 20-h field observation was conducted and written in the field notes.

3 Internet Use Among Migrant Workers

As migrant workers are among the lowest paid groups in the PRD, few can afford to have a personal computer, let alone gain access to the Internet in their accommodations. They also belong to a social group with high mobility (Law and Peng 2006). They not only move between their native villages and their work sites, they also frequently change jobs and move to different factories and different townships. Because they change their accommodations so frequently, they are reluctant to invest in a personal computer. Internet cafés supply them with means of accessing computers and the Internet. In 2005, there were only 314 legal Internet cafés in Shenzhen.[7] By 2006, the number had risen to over 1,000.[8] The wide availability of Internet cafés makes Internet use very convenient for migrant workers.

In the two Internet cafés that are the focus of this study, most clients are young migrant workers. The majority had used the Internet before they migrated to Shenzhen. Some learned how to use the computer and the Internet in middle school or technical school. Some had frequented Internet cafés in their hometown and continued to do so when they migrated to Shenzhen. Usually, they go to the cafés near their factories after they get off work and, most frequently, on weekends and public holidays. There are more than 260 computers in each Internet café. On weekend evenings, it is usually the case that all the computers are occupied and there is a queue for those that come available. The price of Internet access ranges from 2.5 to 4 *yuan* per hour, which is affordable for most migrant workers. A staff member at the Xin Internet café observed,

> Even the client who just has several *yuan* in his pocket can come in and use the Internet. He could get the service that he wants here.

However, the reasonable price of Internet access does not take into account the amount of time spent in cafés. Migrant workers spend an average of 100 *yuan* a

---

[7] Data source: http://city.sz.net.cn/CITY/2005-12/30/content_178521.htm

[8] Ibid.

month on Internet access. In addition to Internet access, some migrant workers spend lots of money on equipment for online games. A 21-year-old man who was addicted to online games spent 800 *yuan* at the Internet café within a month. A 19-year-old worker in an electronic device factory, who updates her blog daily, observed,

> I spent 200 *yuan* to 300 *yuan* per month here [in the Internet café]…. It [using the Internet] has become my main consumption every month. Besides this, I spent little money on other items.

As a medium that integrates many functions (DiMaggio et al. 2001), the Internet provides users with a number of options. The most popular activities among migrant workers in Internet cafés include playing games, watching TV series or movies online, listening to music, chatting on QQ, establishing a personal blog, sending and receiving e-mails, and searching for information. Of these activities, the three most popular are online chatting, playing games, and watching TV or movies. Most of the Internet users do not confine themselves to a single function; they perform a number of activities simultaneously. They chat with their friends on QQ, listen to popular music, and watch TV or movies at the same time. This phenomenon is not difficult to explain: since the price of using the Internet is fixed, users get more value for their money when they access a number of programs.

In contrast to other studies on Internet use (Bimber 2000; Howard et al. 2001; Bonfadelli 2002; Katz et al. 2001), my study found no gender differences in migrant workers' Internet use and no indication that income is a crucial factor. The main difference in Internet use is that between clerical staff/junior management and shop-floor workers. Although they are all migrant workers, the clerical staff and junior management have more education than the shop-floor workers. Most clerical staff and foremen graduated from senior high schools or technical schools, while many shop-floor workers just finished junior high school. Migrant workers with a higher level of education usually are given better jobs and receive more pay. Although the income of clerical staff and junior management is higher than that of shop-floor workers, the former spend much less on Internet access. It seems that clerical staff and junior management are more capable of keeping their Internet spending under control. With the exception of one female member of the clerical staff, all the foremen and clerical staff we interviewed spent less than 100 *yuan* per month on Internet access. Most shop-floor workers, in contrast, spent more than 100 *yuan* per month. This difference could be explained by the differences in the Internet activities of the two groups. For clerical staff and foremen, Internet use is work-oriented and is often confined to data searching and information seeking. As one foreman in an electronic device factory observed,

> I feel the Internet is very useful. If I want to get some data that are related to my work, I can use the Internet to search them. Without the Internet, it would be very hard to get these data….And I often use the Internet to get job information.

A foreman in a toy factory expressed similar views:

> I used to play online games. But now, I feel the Internet is a tool of searching data or information. It is not for other entertainment…. Internet use is for the sake of getting all kinds of information. If I need some data, I will use the Internet to find them.

For shop-floor workers, however, the Internet is a means of entertainment. Several shop-floor workers commented that Internet use made their boring lives interesting and exciting:

> I believe the main function of the Internet is relaxation and entertainment. (A worker in an electronic device factory)
> If I feel bored, I will go to the Internet café. (A young shop-floor worker in a toy factory)

The shop-floor workers mainly use the Internet to chat with friends, play games, and watch movies or TV series. As these activities are time-consuming, their Internet costs are high. This finding reinforces other researches on the effects of education on Internet use: people with higher education use the Internet for instrumental and information-oriented purposes, while those with low levels of education use the Internet "for entertainment reasons" (Bonfadelli 2002: 79). For clerical staff and junior management, the Internet provides a means of increasing employment opportunities and finding work-related information (DiMaggio et al. 2003). They use Internet access to increase their economic welfare and enhance their human capital (DiMaggio et al. 2003). Shop-floor workers, who have less education, see Internet use as a form of recreation that brings relaxation and happiness. However, it is not the intent of this chapter to disparage the use of the Internet for recreational purposes. Relaxation and entertainment play a crucial role in the social lives of migrant workers, which I will discuss in the following sections.

4 Internet Use and Migrant Workers' Social Networks

Social networks are vitally important to migrant workers in the PRD. They provide material aid and emotional support for migrant workers (Lee 1998; Pun 2005). Like mobile phones, which are used to strengthen the social networks of migrant workers (Law and Peng 2006), the Internet provides a way to maintain social ties established in their native province. Online chats are the most popular function of the Internet because they are an economical and convenient way to contact family members and friends. As one worker said,

> I have some good friends who work in Shanghai. Sometimes, they want to give me a call to talk with me. I stop them because making a long-distance call is very expensive. I told them: "If you want to tell me something important, give me a short message. And we make an appointment to get online." We chat on the QQ. For example, one of my friends asked me to buy some goods for him. It is hard for him to explain it clearly by phone. And it is not necessary to give a call. We can talk online.

The cost of making a long-distance call ranges from 0.25 to 0.7 *yuan* per minute, while the average cost of the Internet use is 0.05 *yuan* per minute. When migrant workers want to have a long conversation, the Internet is a prudent choice. In addition, online chats are real-time, visual conversations; they allow for virtual face-to-face interactions. The visual dimension makes communication easier and more intimate. As a worker in a toy factory explained,

> Why do I use the Internet? I have a lot of friends. I often use the QQ to chat with them. It is very convenient for me to contact my family members through the online chat. My parents go to the Internet café near my home. They don't know how to type, but they can use the visual conversation. We use the visual conversation to chat face-to-face.…And it is also easy for me to contact my younger brother and other relatives by the Internet.

Most migrant workers use online chats to strengthen their existing social networks and receive emotional support. However, some young workers are more inclined to establish new virtual communities and expand their social networks by Internet use (Katz et al. 2001; Howard et al. 2001; Wellman 2001). Jane is a 19-year-old graduate of a technical school, who works in an electronic device factory. During the workweek, she goes to Jian Internet café every night and spends 1 h on the Internet. On weekends, she spends more than 4 h at the Internet café. Her primary Internet activity is maintaining her social networks. She divides these networks into two categories—the existing network and the virtual network—and adopts a different approach for each. She contacts family members, former classmates, and existing friends by QQ, but she rarely talks with strangers or net-friends on QQ. She has also created a virtual community for the members of her hometown on the 51 Forum, one of the most popular blog services in mainland China. She named this community, or "group" as she calls it, the "Real Luyang Migrant People" because

> Luyang is my hometown. Actually the group is an association of our locals. All Luyang migrants could join this group. I am the group manager.

As most members of this group are strangers or net friends, Jane insists that the local dialect of Luyang be used in the virtual community, a strategy that eliminates any outsiders:

> [In the group] we don't know each other. But we are locals. We come from the same place: Luyang… If he or she does not come from Luyang, he or she would not join our group. We talk in our local dialect in our group.

Her community has attracted a large number of migrant workers who come from Luyang. In its first 6 months of operation, the community acquired 700 members. As group manager, Jane visits the group every day and spends a lot of time dealing with its concerns. The virtual community has five forums: emotional lives, reading, discussions, pictures, and original stories/photos. Every forum has its own editor, selected by Jane from the group members. In order to regulate activities within the group, Jane has established rules (one requires all editors visit their forums every day). In this virtual community, group members can communicate with each other, post information, discuss daily issues, and share their feelings and experiences. Jane wants to create a space which provides a sense of belonging and social identity for her townspeople. The homepage of the "Real Luyang Migrant People" captures its tone:

> We are so lucky to join this big family. We are proud of being a Luyang people. No matter how hard your life is, our big family always welcomes you. You can get the sense of belonging here.

This virtual community unites migrant workers who are physically scattered. Their interactions in the community strengthen their local identity and the sense of belonging. In the virtual community, they talk freely in their local dialect and share memories of their hometown. These activities lessen their sense of loneliness in an unfamiliar city and provide comfort when they are feeling homesick or unhappy.

For male migrant workers, online games are another important form of recreation, one that, surprisingly, strengthens their friendship in the real world. A young man who works in a garment factory commented,

> I had thought of giving up online games. I did give them up months ago. But after several months, I found that when my friends discussed the new online game, I could not get a word in. I felt isolated. So, I decided to play online game again. When I play games with them, I find we can share so many experiences.

It appears that online chatting and games are not only recreational activities but also important ways for migrant workers to create and maintain their social networks in their new locations. And, if they establish a hometown blog on the web, migrant workers create yet another community—a cyberspace community—where they can develop their social identity in a new way.

5 Internet Use: A Way to Release the Pressures of Work

In the PRD, most migrant workers are employed in the labor-intensive factories established by entrepreneurs from Hong Kong and Taiwan. Heavy workloads, stringent control, and repetitive tasks are the characteristics of these factories (Ong 1991; Lee 1998; Pun 2005). Participants in the study admitted that they had to do overtime work every day. Some were required to work 12 h a day. As many factories adopt so-called "scientific management" of the mass assembly line, it is not considered necessary for migrant workers to understand the whole production process (Braverman 1974; Edwards 1979; Ong 1991; Pun 2005). They are just required to repeat the same bodily movement to fulfill their contribution to the production process (Ong 1991; Pun 2005). Due to the long working hours and the tedious repetition, their work is both exhausting and boring. A young girl employed in an electronic device factory commented,

> My whole life here is working and sleeping. I go to work every day. After I get off work, I go to sleep as I feel so tired. When I wake up, I have to work. My life is so boring.

Given the harsh conditions of their employment, migrant workers need a way to refresh themselves. Internet cafés provide them with an opportunity to relieve the pressures of work:

> Our work is so hard, we need relaxation. I have to do overtime work until ten o'clock at night. I have almost no time to relax. I feel that the pressure of my work is heavy. I go to the Internet café once or twice a week. Sometimes I go to the Internet café at lunch break. Every time I use the Internet in the Internet café, I almost hate to leave. I feel very happy and relax when I use it. (A female shop-floor worker in an electronic factory)

Unlike other entertainment venues, the Internet café not only provides a comfortable and relaxing atmosphere, it also allows its customers to enjoy a relatively independent and personal space. One worker compared the Internet café with the entertainment facilities in his factory:

> There are TV rooms in our factory. We can watch TV there. But, we seldom go to the TV room. The TV room is so crowded that it makes me feel uncomfortable. When I watch TV with so many persons, I feel rather uneasy. You have to behave properly. I cannot lie on the chair when I watch TV with other persons. I feel the atmosphere is solemn, like having a class. But here [the Internet café] is different. I can occupy a computer and use the Internet alone. If I feel tired, I can lie on the chair and use the Internet. It feels like a different world. There is no pressure in this world.

In the factory, the entertainment venue is a public place, shared by many people. Its resources (e.g., the space in the TV room) are limited, and the workers' behavior is constrained. In the Internet café, a personal computer creates a more relaxed private space within a public place. Workers can pursue the activities they enjoy without interruption. Moreover, the Internet café is outside the control of the factory. One male worker maintained that even if their employer provided Internet access in the factory, few workers would be willing to use it because they would still be under the surveillance and the control of management:

> If we use the Internet in the factory, there must be some rules to regulate our use. They will forbid smoking as well as laughing and frolicking when we use the Internet. In the Internet café, it is different. We can do whatever we want, provided that we don't break the computer or make others unhappy.

Migrant workers treat the Internet café as a place where they can breathe the free air and ignore the power hierarchy in the factory regime. A female migrant worker said,

> Our foreman is a tough guy. But when I meet him in the Internet café, I am not afraid of him. He has no right to control me here. He is an Internet user, so am I.

By redefining management as fellow Internet users, migrant workers dismiss work relationships from their leisure activities. They try to erase the effect of the factory hierarchy on their social lives.

Some workers even turn to the Internet to release their negative emotions. Li is a young girl who works on the assembly line in an electronic device factory. She enjoys playing online games in the Internet café. When she feels unhappy or believes that she has been treated unfairly at the factory, she deals with her bad mood by playing online games. Once, at the factory, she put the wrong material into the machine, and nine items had to be destroyed. It was a serious mistake. Her foreman severely criticized her and required her to write a letter of apology and post it on the board. She admitted that she was wrong, but she felt making a public apology was face losing. This punishment depressed her all day:

> I had very bad luck. I made a big mistake. And my whole world becomes dark. I felt that my world had collapsed.

After she got off work, she went to the Internet café and played her favorite online games:

> I put on the earphones and turned the music to the maximum volume. I typed on the keyboard so hard that everyone in the Internet café could hear it…. I indulged myself in the game world. When I play the online game, I feel really comfortable. I release my discontent and anger this way.

The Internet café is a cocoon that helps migrant workers to escape temporarily from their harsh working conditions and the stringent management control. The virtual world provides them with mental relaxation and peace. By viewing the management staff as fellow Internet users, migrant workers temporarily disengage them from the power relations of the factory. Playing online games is also an indirect way for migrant workers to express their discontent with management control in the factory regime.

6 Conclusion

This chapter explores the differentiated uses of the Internet among migrant workers and discusses the impacts of Internet uses on their social lives. My analysis shows that educational attainment is a significant factor in determining the type of Internet use that appeals to migrant workers. It also demonstrates that the differentiated Internet use exists not only between different social groups but also within a particular social group. Various online activities adopted by migrant workers are closely associated with their everyday lives in the city. Low income, harsh working condition and marginalized status in an unfamiliar city, to some extent, explain why migrant workers enjoy online activities and determine why some online activities are more popular than others among migrant workers. Internet use means far more than access to a modern communication technology for migrant workers. They partly experience their lives in the city through Internet use. At the same time, Internet use satisfies the instrumental and expressive needs of migrant workers. Both work-oriented and entertainment-oriented uses of the Internet have a positive influence on migrant workers' jobs and social lives in the PRD. Access to work-related information improves their employment opportunities, while the various means of entertainment provide relaxation. The Internet strengthens the social networks of migrant workers when they are far from home. The online chat, which integrates text, audio, and visual images, is an immediate, vivid, and satisfying mode of communication (DiMaggio et al. 2001, 2003). Migrant workers can enjoy real-time conversations and virtual face-to-face interactions. Internet use helps migrant workers to get "another source of companionship and emotional support" (Haythornthwaite and Wellman 2002: 22).The virtual community is also an important way that migrant workers can keep in contact with others, who are spatially dispersed. The intensive interactions provided by the virtual community (Wellman 2001) provide migrant workers with a strong sense of their local identity and foster a sense of belonging. In addition, the Internet allows migrant workers to relieve the pressures of work and passively challenge the power hierarchy in the modern factory regime. Therefore, Internet use has been deeply integrated into their daily lives.

References

Bimber, B. (2000). Measuring the gender gap on the Internet. *Social Science Quarterly, 81*(3), 868–876.

Bonfadelli, H. (2002). The Internet and knowledge gaps: A theoretical and empirical investigation. *European Journal of Communication, 17*(1), 65–84.

Braverman, H. (1974). *Labour and monopoly capital: The degradation of work in the twentieth century*. New York: Monthly Review Press.

DiMaggio, P., Hargittai, E., Neuman, W. R., & Robinson, J. P. (2001). Social implications of the Internet. *Annual Review of Sociology, 27*(1), 307–336.

DiMaggio, P., Hargittai, E., Celeste, C., & Shafer, S. (2003). From unequal access to differentiated use: A literature review and agenda for research on digital inequality (Working Paper #29). http://www.princeton.edu/~artspol/workpap29.html. Accessed 25 June 2007.

Edwards, R. (1979). *Contested terrain: The transformation of the workplace in the twentieth century*. New York: Basic Books, Inc., Publishers.

Guangdong Statistic Bureau. (2005). *The report of corporation recruitment in the Pearl River Delta*. Guangdong Statistic Information Web: http://www.gdstats.gov.cn/tjnj/table/gaishu_c.htm

Haythornthwaite, C., & Wellman, B. (2002). The Internet in everyday life: An introduction. In B. Wellman & C. Haythornthwaite (Eds.), *The Internet in everyday life* (pp. 3–41). Malden: Blackwell.

Howard, P., Rainie, L., & Jones, S. (2001). Days and nights on the Internet: The impact of a diffusing technology. *American Behavioral Scientist, 45*(3), 383–404.

Katz, J. E., & Rice, R. E. (2002). *Social consequences of Internet use: Access, involvement, and interaction*. Cambridge, MA: The MIT Press.

Katz, J. E., Rice, R. E., & Aspden, P. (2001). The Internet, 1995–2000: Access, civic involvement, and social interaction. *American Behavioral Scientist, 45*(3), 405–419.

Kraut, R., Kiesler, S., Boneva, B., Cummings, J., Helgeson, V., & Crawford, A. (2002). Internet paradox revisited. *Journal of Social Issues, 58*(1), 49–74.

Law, P., & Peng, Y. (2006). The use of mobile phones amongst migrant workers in Southern China. In P. L. Law, L. Fortunati, & S. H. Yang (Eds.), *New technologies in global societies* (pp. 245–258). Singapore: World Scientific Publishing.

Lee, C. K. (1998). *Gender and the South China miracle: Two worlds of factory women*. Berkeley: University of California Press.

Li, Y. H., & Tan, S. (2000). *Migration of rural people and gender*. Beijing: Zhongyuang Peasant Press (in Chinese).

National Bureau of Statistics of China. (2003). *The fifth population census essays* (p. 280). Beijing: China Statistics Press.

Ong, A. (1991). The gender and labour politics of postmodernity. *Annual Review of Anthropology, 20*, 279–309.

Pun, N. (2005). *Made in China: Women factory workers in a global workplace*. Durham: Duke University Press.

Wang, C. G. (2001). The social identification of the new generation of migrant workers and its relationship with city-village merger. *Sociological Research, 3*, 63–76 (in Chinese).

Wellman, B. (2001). Physical place and cyberplace: The rise of personalized networking. *International Journal of Urban and Regional Research, 25*(2), 227–252.

Zhao, S. (2006). Do Internet users have more social ties? A call for differentiated analysis of Internet use. *Journal of Computer-Mediated Communication, 11*(3), 844–862.

Chapter 9
Mobile Phones and the Empowerment of Migrant Workers in Job Search in China's Pearl River Delta

Raymond Ngan and Stephen K. Ma

1 Mobile Phones and Job Mobility

The mobile phone is becoming an increasingly ubiquitous part of everyday life, especially for the younger generation in China. China now boasts one of the largest mobile phone subscriber bases in the world (Glotz et al. 2006). In southern China, increasing numbers of young migrant workers own a mobile phone. To what extent does the use of mobile phones actually promote job change among migrant workers in southern China? What is the possible relationship—do mobile phones act as a push factor or a catalyst to enhance job change? Is it the case that better information via mobile phones makes the job market work better, thus increasing mobility and raising wages? This chapter discusses these research questions based on a survey of 655 young migrant workers that was conducted in China's Pearl River Delta in 2006. Facing the global financial tsunami since October 2008, it is estimated that about 20 million (16.0%) out of the 130 million migrant workers were being unemployed, and the prospects for their continued employment is dim (Macartney 2009). Yet the mobile phone is still continuously used as an active job search device. As for Wang Pan, she was waiting for a phone call from Dongguan City though she had returned to her native county of Wuhan during the Chinese Lunar New Year holidays in January 2009: "I gave my mobile phone number to the factory. I hope the boss will hire me again....At times my native friends used to send me SMS messages informing me updated job vacancies with pay levels to help me in my job search....I do not like staying in my native county as I do not like farming" (Xinhua News Agency 19 December 2008).

China imported its first mobile phone telecom facilities in 1987, and it took a decade for its number of subscribers to jump to ten million (Xinhua News Agency

R. Ngan (✉) • S.K. Ma
Department of Applied Social Studies, City University of Hong Kong,
Hong Kong, China
e-mail: raymond.ngan@cityu.edu.hk

P.-l. Law (ed.), *New Connectivities in China: Virtual, Actual and Local Interactions*,
DOI 10.1007/978-94-007-3910-9_9, © Springer Science+Business Media B.V. 2012

21 October 2005). Since 2000, demand for telephones in mainland China has skyrocketed. Chu and Yang (2006) observed that in July 2003, the total number of users of fixed phones was 240.7 million and that of mobile phones was 239.4 million. By December 2004, when the number of fixed phone users had risen to 312.4 million (an increase of 30%), there were 334.8 million users of mobile phones (an increase of nearly 40%). As for penetration rates, data from the Ministry of Information Industry (2005) show that whereas the penetration rate of fixed phones increased from 17.5 per 100 in 2003 to 24.9 per 100 in 2004 (an increase of 40%), that of mobile phones increased much faster—from 16.2 per 100 in 2003 to 25.9 per 100 in 2004 (a 60% increase). In fact, mobile phones last year contributed almost half of the total income (47.26%) of the telecommunications industry in China (Ministry of Information Industry 2007). The rapid growth of the mobile phone industry in southern China is particularly marked in Guangdong Province, which comes first in the top ten provinces in terms of mobile phone use in China. In 2006, it constituted 35.6% of total sales and income in China's mobile phone industry. Thus, Chu and Yang's (2006) observation that the use of mobile phones is a common pattern among migrant workers in southern China seems to be supported. What has motivated these workers to spend three to four times their monthly income to buy mobile phones?

Law and Peng (2006) found that the extensive use of mobile phones among migrant workers seems to empower them with knowledge about jobs with better working conditions and higher salaries and subsequently increases their job mobility and bargaining power with entrepreneurs. Elsewhere, Chu and Yang (2006) found that the sending and receiving of SMS messages via mobile phones has become a significant communication practice among new migrant workers. What is the relationship between such widespread use of SMS messages and enhanced job mobility, and what are the reasons for job change among migrant workers? Is it the case that the active use of mobile phones supports the hypothesis that, armed with better information about job conditions and market demands, migrant workers with a record of frequent job changes tend to be rewarded with higher wages? Before addressing these research questions and hypothesis, it is necessary to examine the market conditions for the supply and demand of mobile workers in southern China.

2 Market Conditions for Mobile Workers in Southern China

According to the State Council (2006) Research Report on Chinese migrant workers, in 1989, there were already 30 million migrant workers in China. This had been increased to 62 million in 1993 and to 131.8 million in 2006, accounting for 23% of the rural workforce. Seven cities and provinces—Beijing, Tianjin, Shanghai, Zhejiang, Jiangsu, Guangdong, and Fujian—have absorbed 82% of cross-province migrant workers. In Beijing and Zhejiang, about one-third of the total population are migrant workers, while in Shenzhen, it was reported that 12 million of the total 14 million population in 2006 are migrants (State Council, ibid.). Massive economic reforms have been introduced in mainland China since 1978, including the setting up of special economic zones along the Pearl River Delta in Guangdong Province. Many manufacturing industries and factories have been set up, especially in Dongguan region. Entrepreneurs from Hong Kong have been attracted by low labor costs and preferential tax incentives to set up industrial plants in these areas. Each year after Chinese New Year, there is a massive influx of cheap, young laborers from rural counties in the western and inner provinces who are lured to the Pearl River Delta by the good employment opportunities. According to the Guangdong Provincial Annual Digest of Statistics, there were more than 25 million migrant workers in Guangdong Province in 2000, with half of them working in the Pearl River Delta. However, the increase in the supply of migrant workers to southern China appears to have slowed down somewhat, totaling just slightly more than 26.6 million people in May 2007 (Guangdong Department of Labour and Social Security 2000, 2007; Peng 2007). Since the spring of 2004, it has been increasingly reported by the media that the recruitment of cheap migrant workers, especially skilled workers and young female workers, is more difficult (*South China Morning Post*, 26 February 2007). Xinhuanet (2004) found that there were about two million posts unfilled in September 2004, which amounted to about 20% of the total labor force needs in the Pearl River Delta (Xinhuanet 19 December 2008).

The shortage of young migrant workers in southern China has not improved since 2004. In late February 2007, shortly after the Chinese Lunar New Year, the media in Hong Kong and China reported that there was an increasing shortage of migrant workers in Guangdong Province because of its relatively poor working conditions and low wages (*South China Morning Post* 26 February 2007). A survey carried out by the Guangdong Department of Labour and Social Security found that there were 7.3 million job vacancies in the province in 2006, but that only 4.8 million of them had been filled (Guangdong Department of Labour and Social Security 2007). A previous estimate by the Department in 2005 reported a 20% shortfall in the number of workers needed, but observed that the shortfall was particularly acute for labor-intensive industries that offer low pay and poor working conditions (*South China Morning Post* 26 May 2005). The 2006 survey reported that the manufacturing, wholesale and retail, and catering service industries faced the largest shortfall, especially in skilled labor and young female workers aged 18–25. Why is there a mismatch between the supply of and demand for migrant workers in southern China?

At first glance, it seems that the labor supply problem has been aggravated by the central government's declared plan to develop China's inner western regions. Some labor experts have speculated that factories in the Pearl River Delta are losing out to employers in the Yangtze River Delta, notably Shanghai, because the former generally pay less and often flout the labor laws (*South China Morning Post* 9 April 2007). Professor Wei Jie, of Tsinghua University in Beijing, observed that "nowadays, migrant workers have more choices. They can choose to work in Jiangsu, Zhejiang, or Fujian provinces wherever higher salaries and better benefits are offered" (*South China Morning Post* 26 May 2005). Xinhuanet (2004) commented that "during the peak of China's economic boom, over the past 12 years in the Delta, migrant wages had risen only 68 *yuan* or US $8, according to State Council figures. Migrants are actually losing, not gaining, as urban living and food costs are rising." The Christian Science Monitor (2004) noted that "Chinese workers are maturing, and their aspirations are expanding in a manner that implicitly challenges the cheap labor paradigm that makes-up China's comparative advantage." To what extent is the recent labor shortage in the Pearl River Delta, notably Dongguan, related to workers' unhappiness with poor working conditions? What is the possible relationship between mobile telephony and job mobility among migrant workers in southern China? How could the global financial tsunami in the fall 2008 affect their prospects of continued employment, and in what ways could the mobile phone still be able to function as a reliable job search device?

3 Study of Migrant Workers in the Pearl River Delta

The present study was conducted in April 2006 in three industrial towns in Dongguan in the Pearl River Delta: Tangxia, Taiping, and Dongguan City. Successful questionnaire interviews were conducted among 655 migrant workers, 41.98% of whom were in Tangxia, 35.27% in Taiping, and 22.75% in Dongguan City. The 655 respondents were drawn from nine factories in three industrial sectors: metal manufacturing (four factories with 424 respondents, or 64.73% of the total interview sample), plastic products (two factories with 126 respondents, or 19.24% of the sample), and textiles and garments (three factories with 105 respondents, or 16.03% of the total sample). Field interviews were conducted with the help of 14 research and undergraduate students from the Department of Sociology at Peking University, who were supervised by three teachers. The stratified quota sampling method was used, with the strata consisting of industry type (metal, textile and garment, and plastics factories). The survey, in contrast, used the quota sampling method because the intent was to interview all available migrant workers in the factories at the time of the interviews but to exclude local workers born in Guangdong. All of the student interviewers received training and had their completed questionnaires checked by the three supervising teachers.

Why did we choose Dongguan for the collection of our field data? We chose Dongguan because it is considered to be a city of migrants. Indeed, migrant workers make up more than five million (71.43%) of the total population of seven million (Law and Peng 2006). In addition, Guangdong Province has the highest mobile phone penetration rate in China (Ministry of Information Industry 2007).

A total of 655 valid questionnaire interviews were carried out in April 2006. This study can be said to be one of the very few large quantitative surveys of young migrant workers in southern China, as more than half (402, or 61.37%) of the respondents were aged 26 or below, being born after 1980, with 66.26% of them having completed lower secondary education. The sample was made up of 417 male respondents (63.86%) and 236 female respondents (36.14%), 59.24% of whom were single. There were 209 respondents (32.1%) who had worked in Dongguan for

1 year, 162 for 2 years (24.88%), and 132 for 3 years (20.28%), with the remaining 22.74% having worked there for 4 or more years. Most of the respondents came from the inner and western provinces of Jiangxi, Sichuan, Hunan, Hubei, Hainan, and Guangxi.

4 A Record of Job Change Amid the Labor Shortage of Young Migrant Workers

Given the recent shortage of young skilled migrant workers in the Pearl River Delta, our respondents tended to have a record of job change, even those who had only been in the region for a year. Nearly all (79.24%) of the respondents had changed jobs at some point in their working life in southern China. Of these, 39.22% had changed jobs and employers in the year immediately prior to the interview period (i.e., April 2005 to March 2006). The mean number of job changes was 0.38 times per year of working in the Pearl River Delta, and 17.57% of the respondents acknowledged that they changed jobs more frequently than they had 2 years previously. Male respondents tended to change jobs more times than female respondents ($F=8.84$, $p<0.01$).

Of those who had changed jobs, slightly more than half (54.97%) had returned to their native counties but come back to work in the Pearl River Delta after 1 (28.94%) or 3 months (25.27%), and 64.09% of them had also changed industries, for example, from metal manufacturing to the plastics or textile industries. About half (50.58%) of the total number of respondents had left their jobs in Dongguan for a short period but then returned to work in the Pearl River Delta again.

What does this record of job change among migrant workers in the Delta tell us? It seems that underlying the shortage of young, skilled migrant workers in southern China is the shift of these workers from heavy to light industries, or to "factories with a clean working environment," as some of our respondents wrote in their open-ended remarks. They appear to be choosing not only jobs with higher pay levels but also a clean and easy working environment than heavy manual work. The very short period of time that our young respondents spent in their native counties—1–3 months—before returning to work in southern China seems to indicate that although they were farm laborers in their places of origin, they were not keen to work as farmers (Peng 2007; Wang 2003). In China, being a farmer is synonymous with low social status, poor education, surviving with meager income on a small patch of farmland, dressing simple, and living a frugal life (Peng 2007). As school leavers, mostly with a lower secondary level of education, this second wave of young migrant workers (born in the 1980s) tended to be frustrated with farm work but also dislike the dirty working environment in the heavy industries when they first arrived in Dongguan. They tended to move eventually to work in industries with better and cleaner working conditions, especially when the shortage in the labor market allowed them to make such a move. This trend has further aggravated the shortfall in the supply of semiskilled and skilled labor to the heavy industries.

It is not uncommon to see young migrant workers changing their industry and city of employment in and out of Dongguan and the Pearl River Delta areas, but not moving to jobs in the north Yangtze River Delta. It seems that the labor shortage problem in southern China is circumscribed by a second wave of young migrant workers who are basing their employment decisions on the desire to work in more desirable types of industry in better and cleaner working environments with higher wages and to move around the cities of the Pearl River Delta. What has facilitated these job changes?

5 Telecommunications and Job Change Among Migrant Workers

Of the respondents, 417 (64.17%) had a mobile phone, with more men (70.7%) in possession of a mobile phone than female migrant workers (52.8%, $\chi^2=20.95$, $df=1, p<0.001$). Although it appears that the remaining 35.83% of migrant workers were without a mobile phone, it should not be forgotten that 32.1% of the respondents had worked in Dongguan for only 1 year or less, and it was not easy for this group to buy a mobile phone that cost several times their monthly salary. As a great majority of young migrant workers own a mobile phone, it greatly dominates their lives in a broadband society in which fast telecommunications have replaced face-to-face conversation. In addition to using their phones daily (43.96%) or often (33.82%) to make phone calls (83.71%), close to half of the respondents (41.87%) also used them to send and receive SMS messages; 39.45% paid a monthly phone bill of 100 *yuan*. At times, they would also use mobile phone Internet services (36.27%), receive e-mails (21.61%), and search for information on the Internet (35.59%). This shows that the use of mobile phones is widespread among young migrant workers in Dongguan. Similar results has been found by Law and Peng (2006) whose study shows that most migrant workers spend at least 10% of their salary on phone services. Besides, Chu and Yang's study (2006) found that young migrants usually buy expensive phones with features they do not even use, which seems to show that it has become a status symbol, apart from enabling them to strengthen contacts with their boyfriends or girlfriends, especially those who are still living in their home villages.

When telecommunication devices and mobile phones were less ubiquitous one to two decades ago, migrant workers used to rely on the job information that was posted on the notice boards outside of factories and job information centers. The spread of information about jobs, pay levels, and working conditions was less rapid before the days of mobile phones and SMS messages, and that information was less frequently updated. This is no longer the case nowadays because the telecommunication media that are available via mobile phones enable migrant workers to interact actively to determine whether jobs have been filled, whether any friends or coworkers also want to change jobs, and what the track record of prospective employers is with regard to the treatment of their workers, the working conditions, canteen food, housing environment, and overtime work. This is particularly the case when job seekers exchange information with their clansmen, friends, and former colleagues whom they trust even though they are living apart (Law and Peng 2006). With the better and more up-to-date job information that they make available, it appears that mobile phones tend to act as a catalyst to promote and empower migrant workers' quest to change to better jobs and working environments. The cross-tabulation data in Table 9.1 show that among the respondents ($N=606$ people), those with mobile phones changed jobs more often than did those without them ($p<0.001$), especially those who had changed jobs three or more times.

However, the push factor for job change among migrant workers is not the mobile phone. A study carried out by Pun Ngai (2005) found that many migrant workers in China desire to stay in cities rather than return to their rural homes. However, the Chinese government has not resolved the problem of *hukou* (household registration) for them. When this unresolved *hukou* problem is left to entrepreneurs, the cheapest way for them to deal with it is to build workers' dormitories, which has led to the emergence of a "dormitory labor regime" for migrant workers. The problem is that this regime continues to be widely deployed and to generate highly exploitive working conditions and cramped dormitories. This has given rise to the social phenomenon of job change, which represents silent resistance to an exploitive capital-labor relationship. The better, more up-to-date, and faster job information sent to migrant workers by their clansmen, friends, and former coworkers over mobile phones acts as an additional catalyst to promote job change among migrant workers. This is especially the case among the "second-generation" young migrant workers who are the sons and daughters growing up from China's one-child population policy introduced in 1979. Harney (2008) observed that they tended to be more demanding employees: they are pickier about where they work, preferring factories with better facilities and wages. Nevertheless, this phenomenon would not be possible without the shortage of young migrant workers that has existed in the Pearl River Delta since 2004.

6 The Relationship Between Higher Salaries and Mobile Phones

Is it the case that better information via mobile phones makes the job market work better, thus increasing mobility and raising wages? The first part of this hypothesis, that the use of mobile phones serves to promote job mobility, is supported by the findings shown in Table 9.1. This is especially the case among migrant workers who have changed jobs thrice or more. The second part of the hypothesis, that job change among mobile phone owners results in higher monthly salaries, is only true when their first job change led to more money being sent monthly to their native places, as shown and supported by the *t*-test in Table 9.2 ($p<0.01$). However, the data are not statistically significant in terms of the differences in the average monthly salary of workers who do and do not own a mobile phone and the average amount of

Table 9.1 Job change among respondents with mobile phones

Job change	Mobile phone	No mobile phone
No job change	17.1% (66)	27.4% (60)
Changed job once	17.8% (69)	22.4% (49)
Changed job twice	19.95% (77)	20.1% (44)
Changed job three or more times	45.2% (175)	30.1% (66)
Total	100%	100%
Number	387	219

$\chi^2 = 16.751$, $df = 3$, $p < 0.001$

Table 9.2 Salaries and job change among job seekers owning or not owning a mobile phone

	Mobile phone			No mobile phone			Mean	
Respondents	*N*	Mean ($)	SD	*N*	Mean ($)	SD	difference	Sig. (2-tailed)
Average monthly salary	415	566.98	287.47	223	536.59	215.22	30.39	t=0.132 (n.s.)
Average monthly money sent to native place	412	235.97	247.29	219	226.94	216.00	9.038	t=0.649 (n.s.)
Average monthly money sent to native place upon first job change	330	351.95	400.14	160	260.13	275.38	91.82	t=0.009*

*$p < 0.01$

money sent monthly to their native places. It appears that wage differentials are governed by many factors; they are not simply a matter of owning or not owning a mobile phone. Zhao (2003) pointed out that wage differentials are governed by the supply and demand of surplus workers and migrants' age, educational and skill level, and number of years in their current job.

7 Information Sources About New Jobs

Table 9.3 shows the sources of information about new jobs for our respondents. It can be seen that the relative importance of kin as the primary source of information for new jobs dropped from 48.24% for the first job in Dongguan to 37.65% for subsequent new jobs, whereas the importance of receiving SMS messages about jobs from friends and former coworkers and friends made in Dongguan increased from 14.09% to 20.59% and 3.98% to 11.37%, respectively. This shows that there is some relationship between the job search patterns of young migrant workers in Dongguan and the mushrooming of e-Communication in the form of SMS messages

Table 9.3 Information sources for new jobs

Source of job information	% for first job upon arriving in Dongguan (number)	% for new jobs (number)
Kin	48.24 (315)	37.65 (192)
Friends in native place	24.81 (162)	24.12 (123)
SMS job information	14.09 (92)	20.59 (105)
Friends in Dongguan	3.98 (26)	11.37 (58)
Others	8.88 (58)	6.27 (32)
Total	100%	100%
N	653	510

sent by friends and former coworkers. These messages often contain useful, reliable, and up-to-date information about pay levels, working conditions, and living environments in the types of jobs or establishments to which young migrant workers seek to move.

8 A Second Wave of Young Migrant Workers with Different Work Values than Their First-Wave Precursors

Our study has identified a second generation of young migrants from China's rural areas, which is further supported by a similar study carried out by Wang Chunguang (2003). Of our 655 respondents, 61.37% were aged 26 or below, having been born after 1980 when China's one-child population policy became national policy. Wang (*ibid.*) found that this new generation of migrants tends to refuse to recognize its farmer status, which is preset by the existing institutional structure, and imbues itself with more and newer interpretations of identity. However, these migrants lack a clear sense of organizational identity due to the lack of career advancement opportunities in the factories or plants in which they work, and thus they shift from job to job. Because the central government has not resolved to grant *hukou*—or permanent household registration—status to them, they commonly lack a clearly defined status in the existing social structure, which, to a great extent, influences their career, residence, and marriage choices and results in uncertainty over future identity (Peng 2007; Wang 2003). This provides a social context for why our respondents have a record of changing to employment in different types of industries and moving among the cities in the Pearl River Delta to search for desirable wages and better working and living conditions.

Compared with their precursors, this second wave of young migrant workers, born mostly in the 1980s and early 1990s, normally holds different work values. Of our respondents, more than half (59.74%) had changed jobs due to a poor working environment, poor food and lodging, too much overtime, and too little private leisure time. Of these, 12.65% alleged that they did not like the type of industrial plant in which they worked and said they wished to work in a new and cleaner environment.

Table 9.4 Reasons for initially coming to work in Dongguan (N=642)

Reasons	%	Number
To earn money	27.41	176
Fresh graduate eager to "train-up"	24.30	156
Dull work in native place	14.80	95
To learn a new craft for living	11.84	76
To "play" outside the native place	9.50	61
Did not want a farmer's job	4.52	29
To follow their friends to go to Dongguan to work	3.27	21
Others	4.36	28
Total	100	642

This applied especially to those workers who had moved from the metal and plastics industries to work in textile and garment factories. Last but not least, among the reasons for our respondents to change jobs was a lack of staff development and career advancement in the factories or other establishments in which they worked.

Wang (2003) found both economic drive and a quest for a better life to be the motives governing rural to urban migration among migrant workers. Economic drive refers to a move to the city purely to make money to help the family. Quest for a better life means going to the city either for a changed or a more modern life. Our study found a third related factor among young migrant workers: they are fresh junior secondary school graduates eligible to "train-up" (24.30%), in contrast to their precursors who were illiterate or had only a primary level of education. Although "to earn money" (27.41%) was still the major reason given for coming to work in Dongguan, many also reported a desire "to learn a new craft for living" (11.84%) (see Table 9.4).

Table 9.5 presents a different picture and examines the difference between the rhetoric and the reality of our young respondents' expectations in coming to work in Dongguan and their resulting work experiences. There seem to have been other, more prominent reasons for changing jobs than just low wages. It seems that they disliked the poor food and lodging (25%), found there was too much overtime work (20.18%), did not like the industry (as in the case of female workers in the metal industries) (12.65%), experienced poor welfare (no staff development opportunities) (15.06%), and wanted to move to a new job that had more of their friends (9.94%). Thus, as sojourners, the unfavorable working environment in their present factories and a feeling of loneliness in their host society had pushed them to turn on their mobile phones and read SMS job information from their friends and former coworkers. As our young respondents reported in the open-ended questions, they were frustrated that too much regular overtime work in the evenings gave them no time of their own for personal development to study English or computer literacy courses and leisure activities with their friends and clansmen. Forty-five percent of our respondents (N=298) opined that they had participated in the free training

Table 9.5 Reasons for first job change ($N=509$)

Reasons	%	Number
Wages too low	40.08	104
Poor food and lodging	25.00	127
Too much overtime work; no time for rest or leisure	20.18	103
Poor welfare	15.06	77
Poor working environment	14.56	70
Do not like the industry	12.65	64
New jobs have more of my friends	9.94	51

courses offered by their employing factories, but they found that their craft skill levels were still at the basic to elementary level. They did not mind paying money out of their own pockets to study higher craft level courses, more advanced computer literacy courses, and English proficiency courses of longer duration but resented that too much overtime in their workplace was a hindrance to this. The following is an extract from one qualitative comment by a female garment worker in our study: "I pay out of my own pockets to study English because my skills already reached the skilled craftsman levels but my employer said that if I know English, I can be promoted to become a supervisor with twice of my salary. But my company has no such English courses and has much overtime work in the evenings affecting my attendance for English study."

9 Labor Supply Problems in the Pearl River Delta: Underlying Causes

There are social causes for the labor shortage problem in Guangdong Province that go beyond simple supply and demand factors. The first underlying message gleaned from our study is that it is not too difficult to recruit new workers, but it is much more difficult to keep them for very long. For more than two decades, Guangdong's low labor costs have made the Pearl River Delta the manufacturing workshop of the world, making cheap toys, shoes, electrical appliances, computer accessories, and numerous other light industrial products for the world's markets. However, excessive overtime work and poor working conditions are making that workshop increasingly unpleasant for the second wave of new young migrant workers who form the majority labor force in southern China. This new brand of grown-up "single child" (villages and rural areas are entitled to have two children) is less tolerant of unpleasant working environments. A prominent example comes from the London-based *Mail on Sunday* newspaper, which in June 2006 portrayed a leading US plant in Shenzhen as a sweatshop that paid its workers just US $50 a month for 15-h shifts and accommodated them in crowded dormitories (*South China Morning Post* 9 April 2007). As it is difficult to make their present employers offer them better working conditions, young migrant workers are making a silent protest by moving to other factories that offer better conditions. It seems that the factories in the Pearl River Delta, which once highlighted low prices founded on cheap labor, need to repackage themselves as socially responsible, pay better wages, improve staff welfare and working environments, and offer a longer series of computer training or English literacy courses for young migrant workers, rather than running piecemeal to short training courses. It would be timely for local governments and labor departments to offer help and incentives.

The second underlying message is that simply raising the minimum wage would not solve the labor shortage problem. The crux of the problem is not the level of the minimum wage, but the flouting and abuse of labor laws. Although Shenzhen's minimum wage was raised by 17% (from 690 to 810 *yuan* per month) in July 2006, Guo Wanda of the Shenzhen Development Research Centre commented that pay rises alone would not solve the labor shortage problem: "We lack skilled workers, not regular labour. I doubt the minimum salary increases can solve that problem." The media reported that many migrant workers welcomed the minimum wage news, but said the government had to ensure that these pay rises were enforced (*South China Morning Post* 1 June 2006). In fact, abuses of the labor protection laws, such as violations of the law concerning the minimum wage, social security contributions, and limits on overtime, are frequently heard stories among migrants (Chan 2001; Ngai 2005). The matter is further aggravated by a shortage in the steady supply of workers to production units. It is a vicious circle. Factories are getting increasing numbers of orders, but cannot hire enough workers. Their current employees therefore have to do increasing amounts of overtime, their conditions worsen, and, increasingly, they want to leave. An effective solution to improving the high turnover rate among migrant workers appears to be a determination to improve long overtime working hours, poor working conditions, crowded accommodation, staff welfare, and the household registration system.

10 Job Search in the Midst of the Global Financial Tsunami in Fall 2008: Prospects for Young Unskilled Migrant Workers

Although mobile phones seek to empower young migrant workers in the Pearl River Delta areas with instant and updated information on jobs available, pay levels, and working conditions, they are still subject to the economic rule of demand factors for migrant labor. In the golden era for migrant workers during the years 2004–2008 which witnessed a steady shortage of about 20% of the migrant labor force in the Pearl River Delta areas, young migrant workers were able to change jobs at ease checking upon the job search information available in SMS messages or using their mobile phones. Since February 2009, the Chinese government is on its alert that the global financial tsunami is taking a toll on China in slowing down the demand for migrant workers as more and more small- to medium-sized factories in southern China are closing down after the lunar Chinese New Year holidays. Chen Xiwen, who heads the Chinese Communist Party's office on rural policy, found that about 20 million migrant workers, nearly a sixth of the total, lost their jobs in recent months (Johnson and Batson 2009). He further remarked that the total number of migrants seeking work in 2009 will likely be at least 25 million since usually six million to seven million people join the migrant work force each year. To most migrants, returning to work in their native county is not a preferred choice. "I am happy to be reunited with my family, but I am more worried about my job next year. It is cold and I am bored at home," as remarked by 19-year-old Wang Pan who is used to work in an electronics factory in Dongguan city (Xinhua News Agency 19 December 2008).

The Central government announced a four-trillion *yuan* ($400 billion) fiscal stimulus package in November 2008, aiming to help enterprises to tide over their difficulties and to boost economic growth. The extent to which this could be able to help enterprises in avoiding laid-off workers needs time to observe. To most young migrant workers, they continue with their reliance on mobile phones by connecting them either with their employers or friends and clansmen working in the Pearl River Delta areas awaiting return calls for available jobs (Xinhua News Agency 19 December 2008). For those migrants who have been able to find jobs upon their return after the Lunar Chinese New Year holidays since February 2009, they commonly found that the monthly salary is about 15–20% less than the previous year. Although some rural county governments are promoting migrant workers to return to their native counties to start their own business with government loans increased from 20,000 to 50,000 *yuan* and training tickets worth 500 *yuan* (US $73), it is not of good appeal to young migrant workers as these training courses focus on farming, breeding fish, and poultry raising. The "second-generation" young migrants do not like farming or fishing or agricultural jobs. More concerted efforts are needed to help out-of-job young migrants in southern China, notably the use of an Internet job search market linked to their mobile phones, increase the business orders for enterprises and factories in the Pearl River Delta areas by pumping in more money to help small- to medium-sized enterprises to get job orders for the government's infrastructure projects.

11 Conclusion: The Role of e-Communication as a Catalyst for Job Change in a Broadband Society

In a broadband society, e-Communication seeks to promote widespread information flows to a large segment of the population (Srivastava 2006). In Dongguan, the massive use of mobile phones and SMS messages equips migrant workers with quick, reliable, and essential information about job opportunities, notably about jobs with better working conditions, living environments, staff welfare, career advancement opportunities, and pay levels. Messages sent via mobile phones prompt young migrants who are already looking to change jobs to act as "e-Actors" to move jobs together with their friends made in Dongguan. It seems that the e-Communication media have acted as a catalyst for job change among migrant workers. This is supported by our finding that those respondents who had mobile phones changed jobs more frequently than did those without them. Mobile phones, which allow the quick and interactive exchange of information on the latest job market conditions, boost migrants' job information base, increase their job mobility, and empower their bargaining power with factory proprietors. This phenomenon represents a passive protest against the lack of longer training opportunities at higher craft levels, computer literacy, and English proficiency courses, poor working conditions, long hours of overtime work, and low wages among young migrant workers in southern China. Of fundamental concern is the vagrant status of this new generation of young migrant workers, who are excluded from the host society in which they work, but feel uncomfortable with their farmer status in their home towns. Moving to new jobs in different industries, cities, and regions in the Pearl River Delta, they remain uncertain about their future and their social identity. Caught at temporary out-of-job status, some 20–25 million migrant workers are caught in the dilemma of continuing with their job search in southern China or an unwanted return to start their businesses in their native rural counties though they do not like to work as farmers. To those young migrants who do not like to go back to their native counties, they would appear increasingly as peripheral citizens or vagrants in southern China should their unemployed status be prolonged.

Acknowledgments The authors would like to thank the research support of Prof. Yang Shan-hue and the Department of Sociology at Peking University for the help of 14 research and undergraduate students who carried out field interviews in Dongguan. The study was funded by a Strategic Research Grant (Project no.: 7001744) from the City University of Hong Kong. The members of the research team are Dr. Raymond Ngan (ssrangan@cityu.edu.hk), Mr. Stephen Ma (stephen.ma@cityu.edu.hk), and Dr. Pui-lam Law (sspllaw@polyu.edu.hk).

References

Chan, A. (2001). *China's workers under assault: Exploitation and abuse in a globalizing economy.* New York: M. E. Sharpe.

Christian Science Monitor. (2004). Booming China raises workers' hopes and discontent. *Christian Science Monitor, 96*(21), 1–6.

Chu, W., & Yang, S. (2006). Mobile phones and new migrant workers in a South China village. In P. Law, L. Fortunati, & S. Yang (Eds.), *New technologies in global societies* (pp. 221–244). Singapore; Hackensack, N. J.: World Scientific.

Du, J., Liu, X., & Law, P. (2006, September 11). New wave of young migrant workers to work in Dongguan; only 18.2% are just to earn money. *China Youth Daily*, Beijing, p. 2.

Glotz, P., Bertschi, S., & Locke, C. (2006). Mobile phones and mass communications. *Knowledge, Technology & Policy, 19*(2), 3–6.

Guangdong Department of Labour and Social Security. (2000). *Workforce statistics in Guangdong province, 2000. Departmental Reports.*

Guangdong Department of Labour and Social Security. (2007). *Workforce statistics in Guangdong province. 2007*. Departmental Reports. Available at: http://www.gd.lss.gov.cn
Harney, A. (2008). Migrants are China's 'factories without smoke'. http://www.cnn.com/2008/WORLD/asiapcf/02/01/china.migrants/index.html. Accessed 9 Feb 2009.
Johnson, I., & Batson, A. (2009). China's migrants see jobless ranks soar. *The Wall Street Journal*. From http://www.online.wsj.com/article/SB123359977495740100.html. Accessed 2 Sept 2009.
Law, P., & Peng, Y. (2006). The use of mobile phones among migrant workers in southern China. In P. Law, L. Fortunati, & S. Yang (Eds.), *New technologies in global societies* (pp. 245–258). Singapore; Hackensack, N. J.: World Scientific.
Law, P., Fortunati, L., & Yang, S. (Eds.). (2006). *New technologies in global societies*. Singapore; Hackensack, N. J.: World Scientific.
Macartney, J. (2009). China's migrant workers hit by economic pinch as 20 million lose jobs. *Times Online*. From http://www.business.to,espm;ome/cp/il/tol/business/economics/article5638893. Accessed 2 Sept 2009.
Ministry of Information Industry of the People's Republic of China. (2005). Income in telecommunication industries in China. http://www.mii.gov.cn. Accessed 11 June 2007.
Ministry of Information Industry (MII) of the People's Republic of China. (2007). Income in telecommunication industries in China. http://www.mii.gov.cn. Accessed 11 June 2007.
Ministry of Information Industry of the People's Republic of China. (2008). Income in telecommunication industries in China. http://www.mii.gov.cn. Accessed 18 July 2009.
Murphy, R. (2002). *How migrant labor is changing rural China*. Cambridge: Cambridge University Press.
Ngai, P. (2005). *Made in China: Subject, power, and resistance in a global workplace*. Boston: Duke University Press.
Peng, Y. (2007, December 7). *Shortage of migrant workers in Pearl River Delta*. Paper presented at the CSR Link, CSCC Workshop on Migrant Workers' Job Mobility in Pearl River Delta, China. Hong Kong.
South China Morning Post, Hong Kong. (2005, May 26). Talk of tariffs leaves factory floor unfazed, p. A4.
South China Morning Post, Hong Kong. (2005, October 29). Migrants win urban residential status.
South China Morning Post, Hong Kong. (2006, June 1). Minimum wages going up 17% in Shenzhen to ease social unrest, p. A4.
South China Morning Post, Hong Kong. (2007, February 26). Guangdong losing on labour front, p. A6.
South China Morning Post, Hong Kong. (2007, April 9). Bad publicity over sweatshop labour stirs Delta factory chiefs to better workers' lot, p. A5.
Srivastava, L. (2006). Mobile mania, mobile manners. *Knowledge, Technology & Policy, 19*(2), 7–16.
State Council. (2006). *Research report on Chinese migrant workers*. Shuyan Chubanshe.
Wang, C. (2003). *Behavioural selection and motivation for going to the cities of the new generation of rural migrating population*. Beijing: Institute of Sociology, CASS.
Xinhua News Agency. (2005). Mobile better migrant workers' lives. Xinhua News Agency. http://www.china.org.cn/english/null/146181.htm. Reproduced from China Daily 21 October 2005. Accessed 2 Sept 2009.
Xinhua News Agency. (2008). China to help millions of jobless migrant workers. Xinhua News Agency. http://www.china.org.cn/business/news/2008-12/19/content_16975166.htm. Accessed 2 Sept 2009.
Xinhuanet. (2004). *Migrant workers: Urban underclass*. Available at: http://www.Chinatoday.com.cn/English/e2004
Zhao, Z. (2003). *Migration, labour market flexibility, and wage determination in China: A review*. Beijing: China Centre for Economic Research, Peking University.

Chapter 10
Mobile Communication and the Issue of Identity: An Exploratory Study of the Uses of the Camera Phone Among Migrant Workers in Southern China

Pui-lam Law

1 Introduction

The chapter is part of a thesis arguing that the use of the information and communication technologies (ICTs) among the migrant workers has liberated their culturally oppressed personal emotion, has shattered the normative basis of the social structure, has transformed significantly the worker's basic social network, and has reconstructed their social identity. The thesis further argues that the traditional labor capital framework fails to explain the social situation of the workers and that the traditionally framed worker identity should be revisited. If the thesis stands, the state policy on the migrant workers should be reconsidered thoroughly. For practical reasons, the chapter focuses only on the uses of the camera phone among the migrant workers and argues that these uses will contribute to the reconstruction of the worker's identity. And further, if the workers refuse to admit that they themselves are workers and try to redefine their social identity, the state policy, such as the newly promulgated labor contract law, will be meaningless to them.

2 Photography and the Camera Phone

Before analyzing the implication of the use of the camera phone among the migrant workers, a brief discussion of the concepts of photography and the camera phone may be useful. Studies on the camera and photography have been numerous. The camera and photography emerged in the nineteenth century when positivism was thriving; photography was particularly useful, functioning as a mirror with a memory

P.-l. Law (✉)
Department of Applied Social Sciences, The Hong Kong Polytechnic University, Hong Kong, China
e-mail: sspllaw@inet.polyu.edu.hk

P.-l. Law (ed.), *New Connectivities in China: Virtual, Actual and Local Interactions*,

DOI 10.1007/978-94-007-3910-9_10,

(Kracauer 1965). Indeed, the photograph has long been considered a materialized form of memory for communal reminiscing (van Dijck 2008). In particular, studies showed that the photograph was not only an important part of family life, it also served the function of organizing the family memory and constructing the family history and heritage (Chambers 2001; Hirsch 1997; Sontag 1973). This function did not change until the inception of the digital camera and the increasing popularity of the Internet. The photograph sent through the multimedia messaging service (MMS) via the camera phone or posted and shared on the Internet has transformed the function of photography. Photographs now are seldom printed off and archived or compiled in a photo album; they are shared more as experience and less as object (Kindberg et al. 2005). Thus, they are used more often as a visual communication medium and less as a memory tool. Besides, the sharing of photographs has moved from face-to-face to cyberspace, and the traditional ways of oral presentation and collective conversation that build the communal bondage are fading away (Lee 1998). Photographs on the Internet frequently lack a description or caption. The photographs presented have become disenchanted signs and the communication is more decontextualized (Baudrillard 1983). This may shatter one's identity.

Digital photography is more efficacious in shaping one's identity. Users can create their desired memory through the manipulation of photographs (Barthes 1981). Thus, what we think or even what we desire can be built by manipulating the digital image (1981). Even if the users do not doctor the photograph, the easy preview, review, and delete functions on the digital camera allow one to choose the most desired pictures. Photography, particularly digital photography, has largely transformed the visual image, which is in the form of continuity, into a discrete image (Stiegler 2002). Simply put, without the emergence of the camera, we would be in a state of optical unconsciousness, unaware of any picture of the world within a particular slice of time and space. We were living in a continuity with our own visual images and the visual images of others (2002). This forms an understanding, a holistic understanding, of the world and of ourselves. But technology has provided us the ability to suspend our visual image and make it an object image; it can also give us the ability to form our own perception of the world and form our identity. Thus, Stiegler (2002) concluded that "life … is always *already* cinema. The technological synthesis is not a replica…."

Based upon the above understanding, this chapter argues that the use of the camera phone among the migrant workers contributes to a kind of decontextualization and the reconstruction of their social identity.

3 Migrant Workers and the ICT Consumption

Since the early 1980s, when China implemented economic reforms and opened her door to the outside world, leading to the imbalance of economic development between the coastal region and the hinterland, the internal migration from rural to urban areas and from the western and central regions to the eastern coastal region has been continuous. The floating population, predominantly migrant workers, increased sharply from 30 million in the early 1980s to 225.42 million by 2008. This scale of internal migration is unprecedented, and the flow seems incessant unless the stagnant economy of the less-developed region can be invigorated.

In the early 1990s when the economy was thriving, the number of peasants from the hinterland journeying to coastal regions such as Guangdong in the hunt for jobs sharply increased. Most of them led a frugal life so as to save a large part of their wages to use for the welfare of their family, such as building new houses and paying for weddings. Strictly speaking, they did not consider themselves workers even though they were called *dagongzai* or *dagongmei*.[1] These migrant workers regarded working in the city as just a transient period in their lives, and their goal was to return to their home village. These returnees not only brought back economic resources but also "alternate values, life goals, and a new perspective on life" which they had learned and experienced in the cities (Murphy 2002). Our study has shown that the development of new media technologies and the recent rapid diffusion of these technologies have made the less-developed regions more open to developments in the cities lying along the coastal regions. In other words, new forms of city life and new values and ideologies have penetrated the less-developed villages through the new media technologies. Nowadays, through the mediated knowledge, either from the new media or the elder generation, young people are attracted by the idea of life in the city and flock to Guangdong, particularly to the Pearl River Delta, not only to hunt for jobs to improve their standard of living, but also to experience the city life. In general, a large percentage of this cohort of workers does not lead the frugal life of the migrants of the 1990s. They are workers, but they are at the same time active consumers.

Among the workers' consumption practices, the most noteworthy include purchasing new clothes or cosmetics; buying TV sets, MP3 players, and digital cameras; subscribing to mobile phone services; buying the latest model of mobile phone; purchasing Internet services in an Internet café or through the mobile phone; and singing Karaoke songs. In our study, some migrant workers loved to watch TV serials either in their rented apartment or in the common room in the dormitory, and some even went to an Internet café to do so. Some used the mobile phone to send short messages (SMS) to their relatives or friends even when they were working on the assembly line, and some just logged on to QQ (a kind of instant message service which is very popular among the migrant workers and which can be accessed either on the Internet or through the mobile phone) on their mobile, while others went to an Internet café to webcam with their cyber friends. The purchase of the new media technologies and their services has become an important part in their consumption process during the leisure time. In view of all these, we have begun to think that these emerging digitally ambient media technologies will slip the workers into the cyber world insidiously, shattering the normative structure of their face-to-face world,

[1]*Dagongzai* means male migrant worker while *dagongmei* means female migrant worker. According to Lee, *Dagongzai* is a "generic term for workers laboring for the bosses" (Lee 1998). The term *dagong* originated in Guangdong.

decontextualizing them from the workers' life, and in the end redefining the understanding of their identity. We have studied the social consequences of mobile and the Internet use since 2003. Most of our understanding (Law and Peng 2006, 2007, 2008) is on the promising effects of these new media technologies. But recent developments have attracted our attention and we believe that the impact of the new media technologies on the Chinese social structure has been pervasive and deep, particularly on the social lives of migrant workers.

We have been researching the ICTs and migrant workers' social lives since 2003 in the Pearl River Delta. We have interviewed 89 informants; 45 are male and 44 female.[2] Among them, 79 are workers, 9 are factory proprietors and managers, and 1 is an Internet café manager. The informants come from 3 townships in Dongguan city, and 1 is from Shenzhen. Some informants were introduced to us by the factory management, some we met in the Internet café, some were introduced by local villagers, and some by the workers themselves. Our background understanding based on these interviews and observations led us to start looking into the effects of using a camera phone and digital photography in October 2008. We are still at the stage of exploring the issues concerning the reconstruction of migrant workers' identity. Before discussing our thoughts, a few figures on ICT consumption may be useful.

According to the statistics provided by the Ministry of Industry and Information Technology of the People's Republic of China, the number of mobile phone users in China reached 659.8 million (40.8%) by February 2009, with the coastal cities such as Shanghai and Guangzhou taking the lead on penetration rates. Among the migrant workers, the mobile phone penetration rate had gone up to 65% in the Pearl River Delta in 2006 (Ngan and Ma 2008). When we were conducting interviews in 2008 and 2009, we noticed that nearly all the workers had a mobile phone. We also found that the workers viewed a mobile phone as a necessity. In early 2009, we conducted a simple survey in a factory in the Pearl River Delta and of the 90 informants, 60 of them used a camera phone to take photographs and 12 used a digital camera; the others used a traditional film camera or went to a photo shop. Those who used a camera phone were young migrant workers; most of them were born after the 1980s, and some even after the 1990s. In short, the camera phone has become popular among the young migrant workers.

4 Migrant Workers and the Uses of Camera Phone

The issue of camera phone use among the workers came to our attention in October 2008. While we were having dinner in a restaurant, we saw a group of around 30 workers having a farewell dinner with their supervisor. Only two of them had a

[2] Since 2003, the studies have been funded by grants from The Hong Kong Polytechnic University. I wish to thank Liu Xiaojing for arranging the interviews in Tangxia and Humen Towns and for his assistance during the interviews. I would like to express our deep gratitude to Jing Wang, Ke Yang, Yinni Peng, and Chung-tai Cheng for helping me conduct interviews.

digital camera, and others used their camera phone to shoot pictures. As I had my digital camera with me, I volunteered to take a group picture for them. After taking the pictures, I asked for their contact details so that I could send the photographs to them; besides, I also wanted to enlarge the sample size of our study. I found that they did not have any reservations in giving me their contact details, but they were not much interested in keeping the photographs; rather, they just enjoyed reviewing them at the time of being photographed. This seems to be common among other workers. When a waitress and a waiter saw me taking photographs with the workers, they came to me and asked me to take a picture of them. After taking the photograph, they were happy to see it, but they were not interested in having it. I had also observed this behavior when some workers invited my colleague and me to sing karaoke at a restaurant. I noticed that most of the workers had a camera phone, and most of them were China-made. Probably, the price was much cheaper than for imported ones. During karaoke, the workers enjoyed taking photographs with their camera phone; they also enjoyed reviewing the photographs together; they seldom, however, asked to have the photographs sent to them if they were shot by their fellow workmates.

Since then, we have started to explore the social consequences of the use of the camera phone. We have noticed that using photographs as visual communication is common among the migrant workers. Shooting, reviewing, and sharing the pictures are common practice with the camera phone. Photographs were shared together immediately after taking the shot, sent to other mobile phones, or uploaded on to the Internet either through their mobile phone or through computer. When they were sharing the photographs face-to-face, usually they would say, "Oh, have look!" or "Look at this!" and make comments such as, "It is interesting!" or "See how ugly you are!" etc. It was rare that they would give a thick description of the photographs such as telling the story behind them. If they found the photographs interesting, they would exchange them via Bluetooth or infrared device, but they were hardly ever sent as an MMS to their fellow workmates or friends because the data size was large and this would be reflected in the charge. If they wanted to share with a wider network, they uploaded pictures to the QQ website. When we browsed through some of the workers' QQ photo albums, we noticed that the photographs were largely without captions. It seems that they just shared the visual images either together or through the cyber network, and this kind of sharing has become a part of their communication (Van Dijck 2008). Perhaps visual communication has been gradually replacing part of the oral and written communication. Presumably, this would deskill the interpretation ability of the migrant workers (Harris and Taylor 2005).

There are cases of photographs being printed off as a keepsake. A manager in a photo shop told us that such printed photographs were largely of a worker's child or lover. Such photographs would be put in the dormitory or kept in a small handy photo album or wallet; photographs of children were either kept or sent back to family members in their hometown. Some of these photographs were 3 cm × 4 cm, roughly one-eighth of a 3R size photograph. Perhaps, it is more economical and a better size for the wallet. Probably, photographs also have the value of memento.

But even so, among the migrant workers, the practice of printing off the photographs is much less frequent than archiving the photographs in their mobile phone. This is similar in different cohorts of people in different countries (Kindberg et al. 2005). The emergence of the camera phone has transformed the worker's way of keeping mementos.

The type and content of the photographs taken by the migrant workers deserves scrutiny. First, the photographs that we have seen so far from the workers' camera phones or from their QQ website were mainly portraits; they were either the workers themselves or their workmates or friends. Even if the workers were at a farewell dinner, at a birthday party, on an excursion, or wandering in a shopping mall, the pictures they took were predominantly portraits. The background of the portrait would be barely visible, only the people's faces would be in shot. This style of photography is not commonly found, at least among the college students or the white collar workers. Portraits are the focal point of migrant workers' photographs, as was the case when the camera first emerged in the West (Benjamin 1973 [1935]). This is an interesting phenomenon. Perhaps because they are mainly from the economically less-developed hinterland, social modernity is still at the initial stage.

Second, when one looks at the workers' photographs, one can hardly tell that they are migrant workers working in a factory. Very few photographs, either stored on the mobile phone or archived on the website, showed that the workers were wearing factory uniform. One of our informants had only one in her mobile, but that one was taken on the day when she assumed duty; all the others in her mobile showed her in fashionable clothes such as trendy jeans and T-shirts. We have browsed through many photo albums on the workers' website and have also looked through those on their camera phones, and such fashionable presentation was very common among the workers. Although the workers could not manipulate the digital photographs, they could still turn themselves out beautifully.

Third, as the portrait was the focal point of the workers' photographs, few showed the background, so it was difficult to tell where the picture was shot. The obsession with the portrait has made the scene unimportant. But with the older generation of migrant workers in the mid 1990s, photographs of their family and their home were favored. Apparently, this cohort of workers considered the place of their origin important. When one of our informants returned to her work from her hometown, however, there were only photographs of her mother and the children of her relatives, none of her hometown, her village, or her house. Later on, we found that the informant was not happy with her family as her father was a womanizer, and this had worsened the family relationship. She also thought that the family home was old-fashioned. However, photographs of the village where she was working and residing were likewise scarcely found on her mobile despite the fact that the host place is very well developed. She explained to us that she did not find it particularly interesting to shoot the places where she was working. So far as we have interviewed or observed, it is quite common among the workers to lack a photograph of their places of origin and destination.

5 Photography and the Issue of Identity

If one looks through the photographs taken by the migrant workers, one would probably come to the conclusion that most of the photographs lack aesthetic value and a reflective dimension. This is probably true. But the photographs are the product of how the migrant workers see the world and themselves through the camera. The photographs are the suspended moments of the workers' continuity of images (Stiegler 2002). The way they saw things in their hometown, in their factory, and in the host city would probably be more or less the same as the older generation had seen them in the mid-1990s. But for the older generation, photos of their hometown were often important, and they were happy to have photographs of themselves in factory uniform and of the factory in which they worked. They carried all these suspended moments with them. To this older cohort of workers, their hometown was an integral part of their life, and working in the factory was just a transient state. They had suffered enormous hardship when they were working in the host city, but they felt that they still had hope there as they could actualize their goal. The moments of this transient life were therefore also central to their life. They had kept these moments, and they had their history. When we interviewed this cohort of workers in the mid-1990s, they could give a thick description of all those on the photographs. They knew who they were and what their hopes were.

With the young workers, photographs of their hometown were few and far between. Even if they had pictures of their family shot in their house or in their village, very little of the background could be seen. It seems to us that they did not want to capture the moments that they were experiencing when they were in their home village. Since we started the ICT study in 2003, we have noticed that most of the workers we interviewed were not willing to return to their place of origin. There were a number of reasons. Some said that their hometown was economically less-developed, and they believed that there were fewer opportunities there than in the host city; some strongly expressed a dislike for farming; some said that life in the hometown was hopeless; some felt that they would lose their freedom when they returned home. All in all, most of the workers were reluctant to return to their hometown. The idea that the hometown was an integral part of the migrant worker's life was no longer applicable to this cohort of young workers. However, it seems that they were also aware of the fact that they did not belong to the host city. First, the policy of household registration in China prohibited them from registering their household in the host city. They could only obtain a temporary registration. Second, the deep culture of regionalism in China has created a hostile discrimination of the local against the migrant. Third, they were not happy with their work there. In the past, when the economy was flourishing, they changed their jobs frequently. Once they were unhappy with the factory, perhaps because of some minute matter, it was common practice to quit immediately and then move to another factory (Ngan and Ma 2008). Thus, most workers have had the strong feeling of being a sojourner. This is reflected in their photographs. They seldom have photographs of the place where they work. There is no suspended moment there. If photography has the function of visual communication, the lack of photographs of places of origin and destination means that the communication on these is omitted as well. And, if photograph functions as memento, then the memory of these places and their lives there will gradually become blurred. Either case would contribute to the decontextualizing of the workers from the moorings of their lifeworld, which means that the workers would be disembedded from the hometown and the host city. They can hardly be called migrants if they find themselves with nowhere to anchor.

From their photographs, one can hardly call them workers either. This is entirely different from the past cohort. When we were conducting fieldwork in the mid 1990s, we found that the workers wore factory uniform all the time. They seldom took off their uniform even after work. It seemed to us that they were proud of having the uniform; they wanted to tell others that they were different from those who were still wandering around looking for jobs. But when we started our work in 2003, we found that they seldom wore uniform after work. In the street, we could hardly tell who were the migrant workers and who were not. They did not want themselves presented as workers. The representation in the migrant workers' photographs further illuminates the reconstruction of their identity. According to Barthes, photography is the extension of the idea inside our brain of the social and cultural realm (Barthes 1981). The difficulty finding photographs showing that they are workers means that there is scarcely a moment a worker is suspended for the camera. On the contrary, the nonworker representation is abundant either stored in the mobile phone or archived on the Internet. Seemingly, the workers' representation of themselves in a photograph shows that they have the intention, consciously or unconsciously, to reconstruct their social identity. Through the camera, they could suspend the desired moments and store them either on the phone or on the Internet as mementos. They could also transmit them through the camera phone to visually communicate with others their desired moment. The workers have been using the camera to reconstruct their social identity.

If our thought stands, the traditional labor capital framework fails to explain the social situation of this cohort of people. The state policy on migrant workers should also be revised. For example, the newly promulgated labor contract promises them better welfare protection if they are employed by the same factory for 9 years continuously. But so far as we have seen, most of the workers do not stay with the same factory long; they will only find the policy meaningless.

References

Barthes, R. (1981). *Camera lucida: Reflections on photography* (R. Howard, Trans.). New York: Hill and Wang. (1980).

Baudrillard, J. (1983). *Simulations* (P. Foss, P. Batton, & P. Beitchman, Trans.). New York: Semiotext(e).

Benjamin, W. (1973 [1935]) The work of art in the age of mechanical reproduction. In H. Arendt (Ed.), *Illuminations* (pp. 219–245) (H. Zohn, Trans.). London: Fontana.

Chambers, D. (2001). *Representing the family*. London: Sage.

Harris, J., & Taylor, P. (2005). *Digital matters: Theory and culture of the matrix*. London: Routledge.

Hirsch, M. (1997). *Family frames: Photography, narrative, and postmemory*. Cambridge, MA: Harvard University Press.

Kindberg, T., Spasojevic, M., Fleck, R., & Sellen, A. (2005). The ubiquitous camera: An in-depth study of camera phone use. *Pervasive Computing, 4*(2), 42–50.

Kracauer, S. (1965). *Theory of film: The redemption of physical reality*. New York: Galaxy.

Law, P., & Peng, Y. (2006). The use of mobile phones amongst migrant workers in southern China. In P. Law, L. Fortunati, & S. Yang (Eds.), *New technologies in global societies* (pp. 245–258). Singapore: World Scientific Publisher.

Law, P., & Peng, Y. (2007). Cellphones and the social lives of migrant workers in southern China. In R. Pertierra (Ed.), *The social construction and usage of communication technologies: Asian and European experiences* (pp. 126–142). Manila: The University of the Philippines Press.

Law, P., & Peng, Y. (2008). Mobile communication, mobile networks, and mobility: The case of migrant workers in southern China. In J. Katz (Ed.), *Handbook of mobile communication studies* (pp. 55–64). Cambridge, MA: MIT Press.

Lee, C. K. (1998). *Gender and the south China miracle*. Berkeley: University of California Press.

Murphy, R. (2002). *How migrant labor is changing rural China*. Cambridge: Cambridge University Press.

Ngan, R., & Ma, S. (2008). The relationship of mobile telephony to job mobility in China's Pearl River Delta. *Knowledge, Technology and Policy, 21*(2), 55–63.

Sontag, S. (1973). *On photography*. New York: Delta.

Stiegler, B. (2002). The discrete image. In J. Derrida & B. Stiegler (Eds.), *Echographies of television* (pp. 147–163) (J. Bajorek, Trans.). London: Polity.

Van Dijck, J. (2008). Digital photography: Communication, identity, memory. *Visual Communication, 7*(1), 57–76.

Part IV
New Network and New Identification

Chapter 11
Beyond Privileges: New Media and the Issues of Glocalization in China

Boxu Yang

1 Introduction

This chapter explores how the Internet facilitates and induces social change in current Chinese society. How is social order possible in the West? There are many answers in Western sociology. How do social and societal changes occur in the West? The media have played a significant role according to Western media theorists and researchers. How is social order possible in Chinese society? Some answers can be found in Chinese sociology. How do social and societal changes occur in China? It depends upon what literature one reviews. Some may argue that there are revolutions while others insist that there have been no fundamental changes for the past 2,000 years or so. But what was the role of the media in Chinese history? Although answering this question demands knowledge of a considerable amount of research, it is safe to say that media technologies such as papermaking and printing have played very different roles in the West and China. What I wish to argue is that one of the driving forces for the social changes in the West is closely related to the capital or the personification of the capital while the main factors that maintain the dynamics in Chinese society are involved in actively seeking power-related privileges. That is, this chapter contends that the concept of privilege cannot be ignored if we want to have a fuller understanding of the role of media in both the past and current Chinese societies. Accordingly, it is reasonable to explore what the concept of privilege is and how the media are related to it.

Historical fact indicates that China and the West went in very different directions in terms of their political, economic, and social development after the end of feudal society. China ended its feudal system and started its system of autocratic monarchy

B. Yang (✉)
School of Journalism and Communication, Peking University, Beijing, China
e-mail: byang@pku.edu.cn

P.-l. Law (ed.), *New Connectivities in China: Virtual, Actual and Local Interactions*,
DOI 10.1007/978-94-007-3910-9_11,

more than 2,000 years ago while the West started its political journey to democracy in the 1500s. However, there are similarities. For instance, neither road had been traveled before, and each road represented a great adventure. Indeed, the West is still taking great strides along its capitalist road. For instance, how globalization and glocalization will evolve is full of uncertainty. As adventurous as it was, the road that the Chinese traveled was fundamentally an agricultural one which was closely tied with seasonal changes.[1] That is, the traditional Chinese economic activities were highly predictable because agriculture was not commercially oriented. Its social order and change are congruent with or resemble agricultural fluctuations. More specifically, we may say that the social order was maintained through a strict and complicated hierarchical system. The vitality and dynamics of the system were realized by means of delegating power-related privileges to the individuals who functioned according to the needs of the autocratic monarchical system in the Chinese society. Natural resources are limited. The delegation of power-related privilege is zero-sum in nature. The implication is that any imagination or creativity had to be limited to "natural" or nonrevolutionary changes regardless of the media diffused in the pre-Internet era.

2 Agriculture and Social Structure: The Shadow of Nature's Zero-Sum Game in Chinese Society

Western capitalism has many dimensions. Thus far, it is clear that the process of capitalism has moved beyond nature. For instance, the products or commodities produced in capitalist societies are mainly nonnatural goods (e.g., cars) or made of nonnatural materials (e.g., plastics). They are invented or created and have no direct relation to or even nothing to do with products related to the four seasons. In fact, the production and consumption of these nonnatural goods often challenges the natural environment and supports the man-made environment. If we recognize that these nonnatural products are beyond nature, we have reasons to believe that they are not only goods for use and exchange but also symbols that define the Western civilization. One of the dimensions of the civilization is its drive to break down the zero-sum game that is prevalent in agricultural societies. The implication is that the construction of Western societies becomes more and more future-oriented. Values and social structures associated with and treasured in the past become less and less important or even irrelevant in everyday life.

"The medium is the message." Diffusion and implementation of new media often signal societal transformations in the West. Media are significant for they allow their users to be part or to feel to be part of something. That is, they have the potential to invoke or enforce certain social values that are important to social constructions. For instance, the big concepts such as democracy, freedom, science, and technology

[1] There are 24 seasonal indicators in the traditional Chinese calendar.

emerged during the time of printing and electronic mass media. More specifically, media that enable their users to be part of something fundamentally new and great may induce a sense of pride, encouraging users to embrace the man-made environment and form cultural identities during the course of social change. The wider and deeper the real or felt participation is, the higher the degree of pride and identity they may have. The American TV and movie series *Star Trek* may be a good illustration of such values and ideology: "To boldly go where no man has gone before." TV shows can justify economic activities and social costs because they are able to deliver a view of the future in a sense.

A sustainable future depends upon inventions or creations and adventures to a large degree in the West. The endless stream of modern inventions or creations has been driven and selected, among other forces, by the market or the so-called personification of capital. Economically speaking, the significance of those inventions or creations is their ability to empower and encourage their adopters to go beyond the mentality of a nature-oriented zero-sum game. Sociologically, "[t]o boldly go where no man has gone before" may mean "from door to door connectivity to networked individualism" and a "network society" with the assistance of old and new media (Wellman 2002; Castells 2000). While the future may not be ours to see, it seems clear that inventions and creations or innovations are likely accompanied by new lifestyles or ideologies when they are diffused. A new social order or structure is often achieved and maintained through negotiations or dialogues in the process of change although anxieties and risks also exist during the process. The point is that the change and challenge have been part of social life since industrialization in Western societies. The media, old and new, are an indispensable part of the dynamic.

However, the situation is quite different in China. Although what I have described above may be common sense to the Western community, it has significant implications when China and the West are compared in order to discuss the role of new media in social change. The point is that the social structures that emerge in an agricultural society much more likely feature and sustain the zero-sum game because both the power and natural resources are limited. Accordingly, both indigenous and imported inventions and creations are often disciplined or contained whenever it is possible in order to maintain the established structures and institutions. Media technologies are not an exception.

Like many other traditional agricultures, the development of Chinese agriculture was based upon the conditions of nature. What distinguished the Chinese system was perhaps its sophisticated philosophical, social, and political bases. The philosophical arguments indicate that "The Way will not change unless heaven is changed" and "human beings follow the earth, the earth follows heaven, and the heaven follows nature." To the Chinese who initiated the civilization, nature meant the natural world which consists of metal, wood, water, fire, and earth (Cui 2001). The relationship among the five elements is that they move in a sequential cycle. The change of a dynasty is regarded as one particular element that moves to the ruling position while the incumbent and other elements rotate accordingly. This is the political Way interpreted by the official Confucians in accordance with their belief in nature.

The life of a dynasty is regarded as a natural process philosophically. It also becomes natural to construct the social and political structures as supporting systems to the dynasty. Thus, the Emperor is legitimized as the Son of Heaven. His officials become the parents of the ruled or the common people. Among the common people, males act as "Emperors" in their own families or households. The system is proposed and articulated by Confucian scholars. The major structural properties illustrated and endorsed by the Confucians for the system are the three bonds. The three bonds mean that the ruler, the father, and the husband are to be the standards of the ruled, the son, and the wife. Although there were factors that would check the ruler originally, those factors gradually faded away. The three bonds eventually became the three cardinal guides in practice.

The theoretical resources for the system also include the five cardinal virtues:

- Humanity
- Justice
- Propriety (rite)
- Wisdom
- Honor

Confucius-taught humanity, or benevolence, is preeminent among the five virtues. The rest are important only when they help one to realize humanity. That is, humanity has many dimensions. One dimension is "to love others." However, loving others does not mean universal love. There has to be a rank order in terms of blood relationships. The key is to emphasize one's love for one's father (Zhu 1996). This is how the social hierarchy is legitimized at home. Based on the five cardinal virtues, more specific principles were developed by the Confucian scholars. That is, four particular qualities are demanded in daily interactions.

- Loyalty
- Filial piety
- Moral integrity (chastity)
- Justice

To love your father means to show filial piety. Filial piety has two dimensions. One is duty; the other is obedience. The two are important to nurture another quality, loyalty. While loyalty is desired in general, it is particularly important for the emperor or absolute monarch to receive loyalty from his subjects in Chinese society. The rationale is that if a man is a filial son at home, he will likely be a loyal subject when he serves the emperor at court. Similarly, a wife has to serve her husband with moral integrity or chastity in an obedient manner at home. The same quality is demanded of a man when he serves the emperor as a subject. Thus, the social relationships are transformed into power relations. This is the essential structure of the three bonds.

When the structure discussed above manifests itself in daily life, it features rites in terms of the three bonds. During the last Chinese dynasty, the Qing, the court officials often called themselves bondservants in a very demeaning way when they addressed the emperor. A lower official would show similar behavior when he

Fig. 11.1 A middle-rank Qing official in a sedan ("The Last Dynasty: A Collection of Paintings on Qing Dynasty from an English Man." China News Net. http://news.china.com/zh_cn/history/photos/11025921/20050721/12503014_1.html. Accessed 14 July 2008)

interacted with his superior. A county magistrate, the lowest level of administrator in dynastic China, would not be surprised if he was called a parent by the local people. Of course, there were clothing and transportation rites that formed part of the hierarchical structure. A typical example of a rite may be that of transportation for the ruling class (Fig. 11.1).

The above was a painting of a middle-level Qing dynasty official in a sedan in the later eighteenth century. According to the painter, he was on his way to visit the British Envoy on behalf of a top court official. How artistic the work is not important here. The significance is that it offers a vivid picture of the relationship between the ruler and the ruled. Until the Song Dynasty, it was unusual to see Chinese being carried by other people in such a way. But, officials became less and less interested in riding horses or sitting in a carriage that was driven by domestic animals (Lu 1996). They wanted to be carried by other human beings when traveling. Of course, how many people had to carry the sedan depended upon their master's rank. The higher the rank, the more sedan carriers were allowed in terms of rite related state law.

When an official could gain esteem by depriving others of their dignity in the political sphere, a man was able to claim his manhood by putting his wife in a subordinate position and demanding her chastity at home. The roles of a wife were

Fig. 11.2 Kowtow (This is a drawing of a kowtow which shows a son's gratefulness to his parents for giving him life, raising him, and paying his tuition fees. The drawing is by Xie Zhengjun. See "Should a university student kowtow to his parents in order to show his gratefulness". http://news.sohu.com/20070125/n247823894.shtml. Accessed 17 July 2008 (in Chinese))

cooking, bearing children, and tending to her husband and his parents or grandparents. She could not live on her own for she was not allowed to go out and work. In fact, she could not go anywhere on her own with her bound feet. Moreover, later Confucianism dictated she should not remarry if she became a widow, no matter how young or how poor she was. If she did so, she would be regarded as lacking moral integrity. As a well-established Confucian scholar, Cheng, famously put it: "losing chastity is a much more serious matter than starving to death" (Cui 2001).

There were also sophisticated rules on parent-child relationships based upon the rite in Chinese society. In a traditional sense, all the rules may be summarized by a Chinese idiom: "parents are never wrong under heaven." A particular example is that children had to kowtow to their parents either as a form of apology or gratitude on special occasions such as the Chinese New Year[2] (Fig. 11.2).

In short, a parent had complete power over his children, including capital punishment in some cases. Relationships as such were social foundations for imperial China. They were expected to extend to the political sphere, and they did.

Thus, all relations are reduced to power relations to some degree. Power relations are a zero-sum game. Zero-sum games are more likely to occur when human beings compete for limited resources such as land and other natural products. The social structure articulated and advocated by the Confucian scholars is rooted in the traditional agricultural society. The political system supported by them is based upon avoiding a "jungle war" in an either winning or losing world. The solution, as mentioned above, is fixing the winning and losing categories by giving the man, father, and emperor more or less absolute power and stripping away the rights and self

[2] To kowtow is to bump one's head on the ground in obeisance.

identities of others. The problem with this approach is that it works well only when resources are limited. In other words, imagining anything beyond the nature of things would be discouraged for this would endanger the established social and political structures.

But nature can be challenged, as the West made clear from the period of the Enlightenment. In fact, the natural environment has disappeared to a great degree. The man-made environment has become prevalent in today's world. The "unnatural" ideas of science, technology, freedom, democracy, and capitalism are gradually being introduced to China. However, this does not mean that the concept of a zero-sum game becomes less significant. While it is true that many things (both cultural and material) are being introduced to China, these things are carefully selected and disciplined for the purpose of maintaining the status quo. In facing Western challenges, the Chinese have adopted a policy called "Chinese learning as the base, Western learning for application" (see Chen 1994).[3] Chinese learning is mainly based on Confucianism. Western learning is defined as instruments or tools such as science and technology. In other words, the tools are introduced in order to prevent the collapse of the base.

When science and technology are regarded as tools, they become materials because their cultural identity is stripped away. Of course, big Western concepts such as Marxism, freedom, and democracy are also imported. However, they often become an "ism" with Chinese characteristics and are used as weapons or fashionable slogans by interest groups, parties, and commercial establishments for reasons of political struggle. In other words, they are also disciplined by the base of "Chinese learning." The symbolic changes of the past 100 years are clear to see in China: not only have the last emperor and his dynasty gone forever, but a Western material lifestyle has become more and more fashionable. Nevertheless, the social and political structures more or less remain unchanged.

3 Privilege: The Confucian Scholars and Their Imagination

When I say that the social and political structures have not changed much over the past hundred years, I mainly refer to the insistence on "Chinese learning as the base, Western learning for application." Could anything fundamental be changed if the base remained the same? Then, the question becomes: who insisted on the approach and why? It was in the best interests of the Confucian scholars to insist on Chinese learning as the base. One dynasty might be replaced by another, but the Confucian scholars would always be there, ready to serve the new one. Their learning helped any person or group who ruled the country to maintain power and social order.

[3] While Chinese learning is subject to interpretation, it mainly refers to Confucianism.

In return, the Confucian scholars received various privileges as candidates for or participants in officialdom. Yet, privilege was associated with nobles in feudal Europe:

> For much of the last thousand years of European history noble privilege was not merely an expression of social superiority or of economic and political advantage. It was essentially a juridical fact, conferred, or confirmed by royal grant and existing not because of the laxity of the state but because of a legal provision. (Bush 1983: 1)

In general, privileges were enjoyed by the royal family and the officials under the emperor during the past 2,000 years in China. That is, privileges were power-related. Though the wealthy and influential clans received special treatment in terms of taking high office in some dynasties (Lu 1996), they were not nobles in the Western sense. Moreover, being a Confucian scholar and passing the necessary imperial civil examinations gradually became the main means of official selection during China's dynasties.

Being a Confucian scholar and passing the necessary state examination did not mean that one had come from a wealthy and influential clan. Virtually, every man could compete for a government position as long as he had the basic means to support his study and pass the necessary examinations. This was rather significant in terms of social mobility and the dynamic of officialdom in imperial China. That is, when the imperial civil examination became the main system for the selection of officials, the impact of families of traditional importance on social mobility disappeared (Wu 2004a). "To be born into a gentry family did not necessarily insure that one became a scholar or an official in traditional China" (Fei 1968: 17). This means that privileges were frequently redistributed for privileges were power-related in China.

The bureaucratic political system was highly hierarchical in imperial China. The privileges were distributed according to official rank. The officials, retired officials, and Confucian scholars who passed the imperial examination at county level or above were all entitled to privileges in almost every aspect of social, political, economic, and juridical life in the Ming and Qing dynasties, for example. As an old saying puts it: "one man rises to officialdom, then all his dogs and chickens get promoted" (see Fei 1968: 31). The imperial civil examinations were controlled by the state and had nothing to do with truth or knowledge. Confucianism was chosen as the main textbook for the examination because it taught how to serve the emperor and help him to maintain the established order (see Wu 2004b). Under Confucianism, one's highest aspiration is directed to "cultivating oneself, bringing one's own household [family] in order, managing the state as an official, and making the country a stable place as a whole" (Zhu 1996: 1).

In practice, these lofty goals gradually metamorphosed into "studying hard in order to be an official, collecting as much money as possible for personal gain when being an official, buying land and a large house with many servants, marrying a beautiful woman, and building a family that would glorify the ancestor" (Niu 1994).[4] In short, "[t]heir purpose in entering the government was to gain both [political]

[4] The conclusion is based upon study of the Ming dynasty.

immunity and wealth…" (Fei 1968: 31). Indeed, being an official was so important that all the other pursuits were regarded as unimportant. Learning Confucianism and becoming an official were above all trades.

Fei (1968: 65) indicates that "the ideology of the Confucian school represented the point of view best adapted to the Chinese imperial system." At least, we might say that Confucian scholars were responsible for the rites that fundamentally shaped the Chinese political and social structures in the best interests of rulers (Lu 1996; Yan 2004). In return, they were allowed to serve the emperor and were rewarded with the right to enjoy the privileges discussed above. Moreover, all the other schools that flourished during feudal China were dismantled by the emperor (Fei 1968). The Confucian school became the only one left and therefore became the greatest. While all the other schools lost their right to compete to be the greatest, the Confucian schools gradually lost their capacity to be intellectual. The drive to learn Confucianism was mainly born of the desire to become an official entitled to all kinds of privileges. This is the reference point for the imagination of Confucian scholars. That is, the teaching of Confucius was gradually transformed into a tool by its followers.

On the one hand, the system of Chinese imperial bureaucracy was fluid and dynamic because it was open to commoners. On the other hand, it was static and rigid because it was closed to commoners who did not believe in Confucianism. The former makes it fair in terms of social mobility. The latter insures its stability in terms of social change. The imagination of agents or actors was directed or channeled into maintaining the system by means of rewarding officials with privileges. Thus, we understand why the Confucian scholars justified and advocated the men-carried sedan, the status related kowtow, and women's foot-binding. Similarly, it also becomes understandable why science and technology had to be disciplined.

While the Chinese imperial bureaucratic system was officially abolished after the last dynasty, the Qing, the system's spirit lives on to a large degree. Thus, the imagination only goes as far as "the Chinese learning as the base, Western learning for application." From the late nineteenth century to now, "applications" as such have become evident in almost every aspect of Chinese people's life. But "the Chinese learning" seems always able to guide the "application." For instance, China applied aspects of the education systems of the West and former Soviet Union to its education system at different times. However, not only is the idea of a state examination for college entrance almost the same as that of the imperial civil examination but also the purpose of learning in college is similar to that of imperial times.[5] Of course, graduates are now conferred with Western degrees such as B.A., M.A., or Ph.D. and wear Western robes during the graduation ceremony. Imagination and creation seem abundant when it comes to the "application" under the guidance of "Chinese learning." The most recent "applications" include that many officials and successful businessmen become on-the-job graduate students and receive master and doctoral

[5] Personally, I have never heard of any mention of a desire to "make a contribution to knowledge" from the graduate students or colleagues I know during my teaching career as a professor in China.

degrees without actually earning them in mainland China. Moreover, postdoctoral training program has been regarded as part of the status symbol system rather than an academic enterprise.

While imagination and creation as such might be consistent with the essence of Chinese learning and enhance the social and political structures in China, they never really create anything which could be shared by the people. The social order is achieved within the hierarchical social and political system. Capable agents or actors are directed to learn Confucianism, are channeled into officialdom, absorbed by the system through the awarding of privileges, and consumed in the zero-sum game. It is an effective and resilient system for maintaining the established social order. Even in postimperial China, its essential features have never really disappeared. However, the diffusion and implementation of the Internet provide seems to be undermining all that.

4 The Internet and New Adventures for Chinese Society

Maintaining the traditional social order in China partly depends upon (1) the ability to absorb and channel the capable agents or actors into the established hierarchical systems and give them the necessary privileges, (2) the ability to make the ruled think and behave as one, and (3) the belief that a zero-sum game is natural. But the Internet is anything but "natural." Its space is unlimited in a sense and open to anyone and everyone who has access to it. The Internet does not recognize the real world social statuses to a great degree. Its features are best realized in interactions among the users. All of these indicate that what is happening on the Internet is not a zero-sum game. In other words, various opportunities are emerging to the users who enjoy interactions, participations, sharing information, and/or taking new adventures.

The context for those opportunities to emerge is globalization and glocalization. This implies that the two roads traveled by the West and China may be bridged by the economic activities of transnational corporations (TNCs) during the processes of globalization and glocalization. Such economic activities do not merely relocate traditional industries worldwide for profit. They include something beyond the postindustrial economy. When firms such as Google, MSN, and Yahoo landed in China virtually or physically, they were not there to compete for natural resources or cheaper labor to be regarded as a form of cultural imperialism. Similarly, when Chinese Internet companies were listed on the stock exchange in the USA, they were not necessarily there to fight for national interests and were not viewed as an invasion. That is, some of the major TNCs are able to provide the new media-based platforms for the creative industries and enable new social structures to emerge. Creative industries, by definition, have to be creative, interactive, democratic, sharing, and useful (Hartley 2005).[6] Although the new economy is often ignored when we analyze the process of glocalization, it is reasonable to assume that the

[6] China has been promoting the idea of creative industries since 2007.

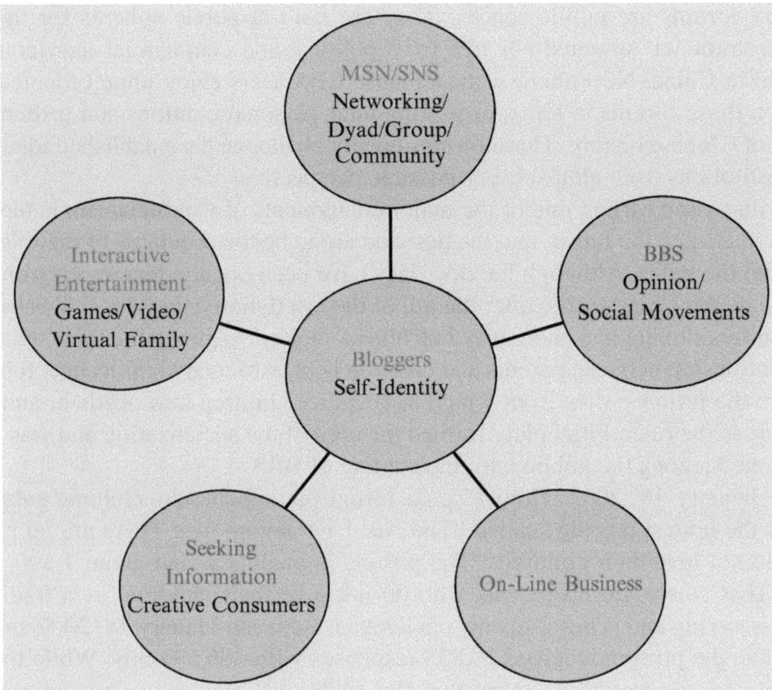

Fig. 11.3 Internet interactions among the users

emergence of new media-based social structures is closely related to the development of creative industries.

If a Chinese new media user is encouraged and/or conditioned to be creative, interactive, democratic, sharing, and useful by the characteristics of new media per se, he or she will have an opportunity to become an agent with strong self-identity.[7] Now, Chinese Internet users seem to be enjoying the interactions from every aspect.

Although Fig. 11.3 is mainly for the purpose of illustration because the categories of users' activities are far from exhaustive, it gives us an idea of the main platforms which have been provided by the creative industries. For reasons of space, only the BBS forums are discussed here. In so doing, I choose postings from the two most influential BBS forums, *Maoyan Kanren* (Looking at Human Beings from a Cat's Eye) and *Tianya Zatan* (Different Views in the Remotest Place on Earth), as examples.[8] *Maoyan Kanren* and *Tianya Zatan* have been ranked as the first and second most visited BBS forums, respectively, in China.[9]

---

[7] There are other factors that have a significant impact on the new media use. For instance, age may be a very important variable.

[8] See http://club2.cat898.com/newbbs/list.asp?boardid=1 for *Maoyan Kanren*, http://cache.tianya.cn/publicforum/articleslist/0/free.shtml for *Tianyan Zatan*.

[9] See http://www.daqi.com/01/01/paihang.html

BBS forums are public spaces. They are not yet public spheres for they are neither "rational" oriented nor free from political and commercial interference in general in China. Nevertheless, the Chinese BBS users enjoy unprecedented freedom on those forums in terms of posting their personal opinions and responses to views of Chinese history. These forums openly challenge the established ideologies and institutions from almost every possible perspective.

As discussed earlier, one of the major components of Confucianism is the three bonds (the ruler, the father, and the husband are to be the standards of the ruled, the son, and the wife). Although the doctrines have been openly denounced from time to time by the Chinese elite after the fall of the last dynasty, they seem always present and functioning in a latent way in Chinese society.[10] Among the three doctrines, the relationship between parents and children is of particular significance, for families are the primary sites from which to socialize children into obedient and loyal subjects to the ruler. Filial piety formed the core of the socialization and was rarely questioned among the public until the coming of BBS.

On January 18, 2007, *Tianya Zatan* forum participant Wu Zhihong submitted one of the hottest postings called "The No. 1 lie among lies: There are no parents who do not love their children." The posting is analytical and about 1,500 words long. That "there are no parents who do not love their children" is a traditional Chinese saying and is never openly challenged. Between January 18, 2007 and July 25, 2008, the posting received 10,833 responses with 366,599 hits. While the vast majority of the responses agree that the traditional saying is very problematic indeed, the participants go one step further to explore and debate why the relationship between parents and children has become as it is in a rational manner.[11] The debate is still continuing and may not end in the near future. But the significance of the debate is clear: parents are no longer sacred and today are subject to discussion and criticism. This has important implications for other sensitive relationships such as the ruler and ruled.

The BBS forum *Maoyan Kanren* seems more concerned with the problems of structures and systems or institutions in China. The participants are able to openly employ the USA as an example to promote freedom and democracy. Sometimes, the ideological orientation of some participants becomes so strong that they are labeled or mislabeled as American government agents.[12] It is true that the debates often get irrational whenever political interests are involved. But they make their contribution to the forum in their own way. That is, everything is subject to discussion, including the rulers and their institutions.

When BBS users start asking questions, power-based privileges become visible, at least on the Internet. Official power or privilege-related corruption and juridical injustice are almost daily topics on the BBS forums. Of course, it is unusual to see

---

[10] Until recently, some local officials called themselves parents of the common people in China.

[11] More detailed data analysis on the posting is underway.

[12] In return, they call the progovernment participants "*wu mao*" which means paid agents.

the men-carried sedan and kowtowing behavior in public in current China. But it is easy to see that the automobile, not the pedestrian, has the right of way on the streets in China's cities. To some extent, we may say that it is the automobile technology that has made the men-carried sedan disappear. But the technology alone cannot change the privilege mentality. A posting on *Maoyan Kanren* clarifies the problem. The posting appeared on January 21, 2008 and was entitled "The whole society is the loser (Hai shi Guo Ke 2008)." It examines a whole range of occupations and warns that if the majority of people (officials, teachers, businessmen, and common people) continue to work for personal gain, the whole of society will collapse. This particular posting has received 433 responses. The discussion is enlightening. Postings are trying to nourish the idea of civility and citizenship in China. However, political interests prevent the BBS users from going much further.

Not only are the topics on the BBS forums from the real world, but the popularity of the topics and the debates reflects real life in Chinese society. That is, the BBS forums also fall under the shadow of a zero-sum game to some degree. This is crystallized whenever the posting argues for a return to the Maoist system or uses American democracy as a slogan to attack the current regime. Understandably, insults and abuse are often exchanged during the debate because the posting opens the old wounds that reflect modern Chinese history. The point is that imagination is lost in the fight. But taken as a whole, the BBS forums do provide a platform for communication and interaction. When different views or facts are presented, it is for the BBS users to judge which one is right or wrong and decide to respond or ignore. This is particularly important to the younger generations who are not familiar with modern history such as the Great Cultural Revolution.

The Chinese BBS forums, like other Internet platforms, are inherently interactive. Active involvement with the discussion and making a quality contribution to the debate may lead to meeting other users off-line and lasting friendships. This is also particularly important to the younger generation who are concerned with social capital and mobility. Social mobility is emphasized because it is privilege-related in Chinese society. It is part of a zero-sum game and closely associated with social capital or social ties (Bian 1994; Bian and Ang 1997). While weak ties are significant in terms of getting a job in the USA (Granovetter 1973), strong ties are more important in China (Bian 1997).[13] However, the strong ties mainly refer to the relationship with family or household members who can be trusted and obligated. In other words, family has a much stronger impact on the social mobility of the next generation in China than in the USA. Hence, we have reason to believe that the Internet communication in China may also have a different impact on developing social ties beyond the closed ones and increasing social capital through interaction with strangers on the Internet. Studies reported in the USA indicate that "the Internet

[13] According to Bian (1997), personal networks are employed to influence authorities who in turn assign jobs as favors to their contacts, which is a type of unauthorized activity facilitated by strong ties in China. The study was conducted before the diffusion of the Internet in urban China.

is increasing social capital" (Quan-Haase and Wellman 2002: 319), although the linkage may not be direct (Katz and Rice 2002).

The Chinese BBS forums are just one application of the Internet. Somehow, it is hard to imagine it could be used as a social platform.[14] But if the users are able to show mutual appreciation of each other's postings or responses on the BBS, they will have a solid base from which to build friendships when they meet off-line. An interview with a former graduate student of Peking University reveals that she first got to know her best friends on the BBS. After they met, they exchanged mobile phone, MSN, and QQ numbers. After that, they maintained contact and organized off-line activities such as group journeys through all the means of new media. Interviews conducted with her best friends also indicate similar networking experiences. Although the social capital or social ties are not directly increased or developed by posting per se, they are extremely valuable. The former student interviewed currently works in Guangzhou which is far away from her home town as well as the cities in which she completed her undergraduate and graduate programs. The "strangers" she saw on the BBS have become her best friends in the same city, and they have provided all kinds of support to each other when necessary.

In sum, what may be expected from the Internet for the Chinese users is a deconstruction of the traditional social structure and political ideology. As such, the ability to interact with strangers and communicate effectively may be increased. This is the first step to gaining the civility and citizenship that are crucial for the creative industries to flourish. In return, the users' contributions are appreciated and possibly transformed into social capital.

5 Concluding Remarks

The Chinese society is built on the system of zero-sum games and power-related privilege. Individuals know very well how to interact with family members and superiors or subordinates. However, the ability to communicate and share with strangers on an equal footing is very weak. Moreover, although the state examination for university entrance has been regarded as fair in terms of social mobility, it very much discourages imagination and creativity. This is in sharp contrast to the Western experience.

The diffusion of the Internet provides an opportunity for the two societies to truly meet each other during the process of globalization and glocalization. The Internet-based process of glocalization has great implications for Chinese society. If the family ties are reexamined and reconstructed, if the "strangers" are able to be included in the social networks, if the hierarchical system is challenged, and if the power-related privileges are stripped away, true change will come to Chinese society.

[14] Many of the BBSs provide a private channel of communication for their users in China.

References

Bian, Y. (1994). *Work and inequality in urban China.* Albany: The State University of New York Press.

Bian, Y. (1997). Bring strong ties back in: Indirect connection, bridge, and job searching in China. *American Sociological Review, 62*(3), 366–385.

Bian, Y., & Ang, S. (1997). Guanxi networks and job mobility in China and Singapore. *Social Forces, 75*, 981–1006.

Bush, M. (1983). *Noble privilege.* Manchester: Manchester University Press.

Castells, M. (2000). *The rise of the network society* (2nd ed.). Oxford: Blackwell.

Chen, M. (1994). Chinese learning as the base, Western learning for application: In addition to the enlightenment and saving the nation from subjugation-modern indications of Chinese culture. In *Original Way*, No. 1, Beijing: Chinese Academy of Social Science Press.

Cui, D. (2001). *An introduction and discussion on Confucian studies.* Beijing: The People Press (in Chinese).

Fei, H. (1968). *China's gentry.* Chicago/London: The University of Chicago Press.

Granovetter, M. (1973). The strength of weak ties. *The American Journal of Sociology, 78*, 1360–1380.

Hai shi Guo Ke. (2008). The whole society is the loser. http://club2.cat898.com/newbbs/dispbbs.asp?boardid=1&id=2044384. Accessed 25 July 2008.

Hartley, J. (2005). Creative industries. In J. Hartley (Ed.), *Creative industries* (pp. 1–40). Oxford: Blackwell Publishing.

Katz, J. E., & Rice, R. E. (2002). Syntopia: Access, civic involvement and social interaction on the net. In B. Wellman & C. Haythornthwaite (Eds.), *The Internet in everyday life* (pp. 114–138). Oxford: Blackwell.

Lu, S. (1996). *Lu's general history of China.* Shanghai: Huadong Normal University Press (in Chinese).

Niu, J. (1994). The social significance reflected by the Ming scholar groups. *Monthly Journal of History, 2*, 32 (in Chinese).

Quan-Haase, A., & Wellman, B. (2002). Capitalizing on the net: Social contact, civic engagement, and sense of community. In B. Wellman & C. Haythornthwaite (Eds.), *The Internet in everyday life* (pp. 291–324). Oxford: Blackwell.

Wellman, B. (2002). Little boxes, glocalization, and networked individualism. In M. Tanabe, P. Besselaar, & T. Ishida (Eds.), *Digital cities II: Computational and sociological approaches* (pp. 10–25). Berlin: Springer.

Wu, Z. (2004a). Introduction. In Z. Wu (Ed.), *A study of bureaucratic political system in Ancient China* (pp. 19–86). Beijing: Peking University Press (in Chinese).

Wu, Z. (Ed.). (2004b). *A study of bureaucratic political system in Ancient China.* Beijing: Peking University Press (in Chinese).

Wu, Z. (2007). The no. 1 Lie among all the lies: There are no parents who do not love their children. http://cache.tianya.cn/publicforum/content/free/1/854388.shtml. Accessed 25 July 2008.

Yan, B. (2004). The bureaucratic political system in the earlier stages of empires: Qin and Han. In Z. Wu (Ed.), *A study of bureaucratic political system in ancient China* (pp. 19–86). Beijing: Peking University Press (in Chinese).

Zhu, X. (1996). *Variorum edition of selected sentences and chapters of the four books.* Jinan: Qilu Press (in Chinese).

Chapter 12
The Principled Machine: A Sociopolitical Inquiry of Mobile Voting in Chinese Society

Chung-tai Cheng

1 Introduction

One of the major promises of mobile technology is that it will increase unity in communications. The recent development of the mobile phone is playing a revolutionary role interconnecting mobile telephony with the press, radio, television, camera, and desktop computer, which is in turn contributing to new communication structures resulting in unexpected social consequences (Pertierra 2005). By enhancing communications, mobile technology is expected to promote social and political participation. In principle, the mobile does offer a technological precondition for such participation, but the understanding and practice of different and very specific societies about this ideal are another issue.

With the mobile phone being the most-diffused technology in China, Chinese people can enjoy not only greater accessibility and flexibility of interpersonal communications but also increased opportunities in social and political participation. For instance, Chinese people organized popular protests against the French international hypermarket chain Carrefour S.A. (which resulted in mass boycotting) by SMS and by communicating online in the spring of 2008 because the company was accused of supporting the Dalai Lama and pro-Tibetan independence groups (Cheng 2009). Apart from mobilizing the public to participate in social movements, some institutions, companies, and organizations started to make use of the mobile as a means of conducting polls and elections, such as by voting for the most important Chinese character in New China, the most beautiful tourist scene, the best representative in class, etc. Hence, the mobile has become a previously nonexistent platform for e-participation in Chinese society. In particular, the increasing influence of this phenomenon has attracted more and more government attention – and this probably

C.-t. Cheng (✉)
Department of Applied Social Sciences, The Hong Kong Polytechnic University, Hong Kong, China
e-mail: sscct@inet.polyu.edu.hk

P.-l. Law (ed.), *New Connectivities in China: Virtual, Actual and Local Interactions*, DOI 10.1007/978-94-007-3910-9_12, © Springer Science+Business Media B.V. 2012 has encouraged the government to change its attitude and working style during the policy-making process, particularly in media policy and related regulations. The key question here is what the sociopolitical implications of mobile voting in the eyes of different parties are if Chinese society is understood as differentiated and graded social relationships, or a "differentiated mode of association" (Fei 1992), in which social obligations within the network are often characterized in familial terms.

By drawing on an example of a Chinese karaoke contest, the 2005 show *Super Girls* (renamed *Happy Girls* in 2009), this chapter attempts to explore the social consequences of mobile voting in Chinese society. In the first section, the chapter provides a brief outline of *Super Girls* in order to highlight how the TV show utilizes mobile technology to enhance its popularity and audience ratings and to show the interaction between different parties. In the second section, the chapter presents two sides of a debate portraying mobile voting as first a democratizing innovation and second as a profiting one. The debate highlights indigenous Chinese understanding on elections. In the last section, the chapter will discuss how mobile voting may be considered as a potential threat to Chinese society, especially in moral, political, and official terms. In the conclusion, the chapter will attempt to demonstrate that the Chinese government is trying to define and restrain social media as a principled machine with exemplary functions in reaction to the power it gives to the people, and this may contribute to a new understanding of sociopolitical negotiation in the Chinese context.

2 From Super Girls to Happy Girls

Super Girls (*Chaoji Nu Sheng*, or literally "Super Female Voice") was an annual national singing contest in China in the vein of the US show *American Idol*. The show ran for three consecutive seasons from 2004 to 2006, and ran again in 2009. The series was organized by Hunan Satellite Television and was open to any female contestant regardless of her origin, appearance, personal record, or singing talent. The almost unrestricted auditions for the 2005 series attracted more than 120,000 applicants from all over the country, aged 4–89 years old (Qing 2005). Competitors went through preliminary selection rounds which were held in the five provinces of Hunan, Sichuan, Guangdong, Henan, and Zhejiang. The remaining singers would engage in a weekly broadcast knockout competition, in which the two contestants with the lowest judge votes and audience SMS vote would leave the show. TV viewers were also encouraged to watch each of the one-to-one matches and vote for their favorite singers by telephone and SMS, resulting in subsequent eliminations until only three contestants were left to compete to win the series.

The show drew a massive audience with more than 280 million viewers tuning in to the 2005 season finale, surpassing even China Central Television's annual Spring Festival Evening Gala as the most viewed TV program of the year (Goldkorn 2005). However, along with great popularity came greater scrutiny, and certain government organizations began to worry that the show was "poisoning youth," promoted vulgarity, and encouraged youngsters to seek instant celebrity (Martinsen 2005). The State Administration of Radio, Film, and Television (SARFT) in China therefore released a series of measures to regulate TV entertainment programs, particularly talent shows.[1] Afterward, the competition was relaunched for the 2009 season, with the name changed from *Super Girls* to *Happy Girls*. In addition, the program organizer had to agree to certain conditions imposed by SARFT designed to limit the "sensationalization" of the show. For example, the competition could now only last for 2 months, and each episode had to be aired after 10:30 p.m.; the program could only include "healthy and ethically inspiring" songs; the producers had to limit gossip about the contestants and scenes of supporters (including participants' family members, relatives, and friends) screaming and wailing; judges had to hold themselves with some decorum and be authoritative and observe common courtesy; and televised scenes of out-of-control fans were to be avoided (Ye 2009). Last but not least, all forms of public voting for contestants, including text messaging and online polls, were prohibited (China View 2009; Shanghaiist 2009).

3 Democratizing or Profiting?

The popularity and success of *Super Girls* not only generated great fortune for its stakeholders but also great debate among the ordinary people about the implications for Chinese society of a mobile or online voting system. We can find in this debate two main points of view. One side maintained that the show represented an example of superficiality in contemporary Chinese society, in which the combination of audience participation and mobile technology was only one more moneymaking tool available to commercial entertainment enterprises (Zhou 2005). In particular, the design of the SMS voting method was in serious doubt. According to the rules of the show, each mobile number was able to vote a maximum of 15 times a week, each vote costing 0.5–3 *yuan* (the usual cost of an SMS message is 10 cents). Since China's mobile service providers had not required users to register for the service using their real names, a single person could buy multiple SIM cards to vote for his/her favorite singers, with the only requirement being able to pay. Lower-income masses would therefore carefully consider the cost of participation, whereas higher-income viewers could easily buy more phone cards and tickets and use their money to influence the result of the competition.[2] As a public intellectual in Shanghai commented,

---

[1] The State Administration of Radio, Film, and Television (SARFT) is an executive branch under the State Council of the PRC that is mainly responsible for controlling all levels of media enterprise in China, both state-owned and nonbusiness, and also censoring any materials that might be damaging to the Chinese government or cultural standards.

[2] Though the issue of the registration requirements for mobile phone subscribers in China has been discussed for many years, prepaid SIM cards are still dominant and widely available. This means mobile phone users can have several numbers because the cost of applying for a new number is only around 50–100 *yuan*.

"The penetration of money into the voting is a subversion of the democratic spirit. It does not represent the will of the people; [it] only represents the will of money… Behind the Super Girls, there is an invisible hand – it is the hand of money, it is the hand of power…the hand that stole, toyed [with], and manipulated public opinion to realize its quick commercial profits" (Xu 2005). Netizens also performed a postmortem on the show in Sunday's *Legal Mirror* and commented, "'Supergirl' is a money game. Income from mobile phone SMS topped 30 million *yuan* (US$ 3.7 million)…Experts have calculated the 'Supergirl' brand by itself is worth at least 100 million *yuan* (US$ 12.3 million). When a 'Super Voice Girl' can bring in this sort of cash, how can we not submit?" (Danwei 2005).

However, another side of the debate suggested that *Super Girls* was a milestone in China's entertainment industry, proof that tele-entertainment was no longer the preserve of the rich and the elite but also of the masses, an example of grass-rooted democratic expression. In addition to the unrestricted application requirements, viewers were able to participate in the judging process by sending an SMS or being online to vote for their favorite contestants. Though the final result of the competition was determined by three means – professional judges, audience judges, and the number of votes – viewers had a certain amount of power to define the progress of the show by putting pressure on the organizers and judges so that they had to consider fairness, justice, and the consequences to public opinion. Thus, some culture and media experts praised this practice for blazing "a trail for grass-rooted cultural democracy" and offering the Chinese public a valuable opportunity to practice and learn from an alternative rehearsal of democratic elections (Zjuzhanglei 2005). Like the 2005 contest finale, the top three contestants alone drew more than eight million message votes so that it was considered as one of the largest "democratic" voting exercises in China (Yardley 2005). As a Chinese commentator who is famously critical of popular culture maintained (Sohu 2005):

> *Super Girls* pioneers a cultural democratic election through 'thumb voting'…the remote control device is just in your hand. If you want to watch the program, just turn it on; otherwise hit the off button and leave. This is the 'democracy of viewership' which reflects the will of the people…The *Super Girls* gives us a lesson that public opinion can generate powerful force when the media can grasp it. In my view, it is the most important power of production in 21st century China.

Although the practice of mobile voting may be considered as a means of profit making, it is reasonable to suggest that the popularity of this show represents an important step for Chinese society, in which the advancement of communication technology and the combination of old and new media promote a participatory experience that is creating brand new notions of viewership, ownership, and storytelling. Undoubtedly, having the practice of voting does not simply mean that the society can be more democratic than before as people can still make use of legal gaps to buy votes or cheat during the voting process. The chapter, however, wants to discuss how SMS voting enables significant changes in the role of TV viewers, from spectators to social audience before exaggerating the weaknesses of voting system.

As Kuhn (1992: 301–311) suggests, "spectators" are individuals who engage with a TV show but who may or may not do so with any sense of belonging to a larger collective of viewers, whereas "social audience" can be understood as a collective of people, either real or fictive, coming together to watch and participate in a show, guided in part by the work of the television industry. The notion of the distinction between the social audience and the spectator is extremely important for understanding the social meaning of mobile voting for Chinese people because its vision is related to individual freedom and emotional expression, which is usually paid little attention in traditional Chinese cultures. SMS voting provides a more direct, more convenient, and faster way for Chinese people to express their private emotion, preferences, and opinions. This simplified voting process can be completed in a few minutes by pressing a few buttons and moving a thumb. Due to the convenience and low cost of participation, SMS voting not only enhances people's desire to participate in something that they are interested in but also contributes to a sense of belonging and accomplishment that the Chinese had never experienced before. The foundational issue here is how such an element of "e-sociality" can be understood in the Chinese context. Though cell phones are usually considered as individual communication devices, it does not mean that we can simply regard the phones itself as a handy tool and understand the meaning of SMS voting in terms of the concepts of individual autonomy and expedient rationality. During the contest, some news channels reported that there were companies providing professional fake fans (or "fans-for-hire") to help contestants boost their popularity among voters (China.cn 2005). One contestant's mother even revealed that she owed one such company 90,000 *yuan* and that this was a disaster for her family (Huaxia 2006). This shows that practices of SMS voting are embedded in a complex structure of Chinese social relations, representing a pragmatic and unprincipled logic: if there lacks *guanxi*, or a network of mutual support, the most efficient and effective way to win the game is to buy votes from the market.

4 The Principled Machine and Exemplary Traditions

To the ordinary Chinese people, *Super Girls* is not merely an entertainment program but a larger social event. By taking part in an SMS vote, not only are they encouraged and able to participate in the show but they can also show individual preferences and share personal opinions and responses through a multimedia interface. The Internet, especially BBS forums and instant messaging programs, enable TV viewers to share and discuss a range of experiences of and opinions on such tele-participation and related topics. As Ross (2008: 46) suggests, "The Internet provides an immediacy and sense of ease for viewers who go online, creating 'something more' of something that is familiar – fandom." The counting of votes and the outcome of the competition could have been the end of the TV show, but it was in fact the beginning of *another* show. The social audience not only paid attention to and discussed the details of the contest on the Internet but they also personally experienced and repeatedly reflected on and discussed their own choices with others. Though concern about personal preferences and interests motivated them to participate, Chinese people were able to experience and learn from a collective sense of social (cyber) involvement, and that might eventually cultivate an air of online social participation as a lifestyle. In other words, mobile communication seems to create a room for individuals to express their private emotions in a manner which may not be culturally or morally acceptable and might even contribute to the undermining of the foundation of Chinese social institutions and even the dynamic of the traditional power structure between Chinese officials and the ordinary people. In order to address how the Chinese government interprets and responds to the emerging "sociality" in digital space and shed some light on its rationale for prohibiting mobile voting in entertainment shows, we must explore the issue of power and its relationship to the combination of tele-participation and mobile communication, and especially how the sense of ownership translates to the issue of discourse and representation to the Chinese government.

The idea of "voting," or more concretely "democratic elections," has been prominent in Chinese discussions regarding China's governance and possible democratization, but there is a complicated understanding of the meanings of this term in the Chinese context. In Chinese culture, the focus of the problems of governance and the possibilities of democratization is on sustaining social integration and community well-being (Fei 1968). Though the meaning of community may vary considerably from family to clan to nation, it represents the essence of Chinese society building, which is not viewed in terms of treating someone as an individual on equal terms, but rather according to relationships between persons of oneself. As Hsu (1981) maintains, the Chinese notion of "self" is understood as being different from the rational, self-conscious, and autonomous Western concept of "self"; it is a "relational self" which tends to be socially or psychologically dependent on others, for this "situation-centered" individual is tied closer to his world and his fellow men. By emphasizing culture and the shared practices of the society as the source of direction, guidance, meaning, and value in social life, "individuals are considered as players in a collectively orchestrated symphony of social exchange" in the community and society (Rosenberg 2006), that is, one must participate in and learn from the experience of interacting and communicating in person in order to learn how to behave in accordance with the time and the place. In contrast to the public-private divide in Western society, influence and control mostly flow from the public sphere to the private one in China (Kinnvall 1995). Therefore, the traditional belief is that "power can be restrained and that the good society will develop through morality and strict standards based on objective norms," and so correct behavior is always linked and evaluated with close reference to the objectivity of social norms and exemplary models (Bakken 2000).

Based upon this brief understanding of the Chinese traditional culture of governance, we can see that the Western idea of "elections," which is anchored by a focus on the nature of the individual citizen, in principle, is completely different and even contradicts the normative emphasis of China's top-down and central management approach to the ideal state of being a perfect, good, and stable society. As Metzger (1998: 30) maintains,

> Modern Chinese political thought has not yet turned toward an un-utopian, bottom-up approach. Based on the traditional optimism about political practicability, it still reflects the paradigm of a morally and intellectually enlightened elite working with a corrigible political center morally to transform society, instead of emphasizing the organizational efforts of free but fallible citizens forming a civil society with which to monitor an incorrigible political center.

An interesting article by Jin Rong (1989; translated by Bakken 2000) discusses the futility of elections in China and maintains, "Some people simply vote according to their own likes or dislikes, whether they are related to people, whether they have connections or *guanxi* with them. If we use elections instead of evaluation, the unity and stability of the work-units will disappear...the ballot box is not based on a principle of goodness or perfection; it becomes unprincipled and without any moorings in the correct and principled social norm." It is interesting to note that similar comments were also delivered during the singing contest. China Daily's cultural critic Zhu Dake asks, "How come an imitation of a democratic system ends up selecting a singer who has the least ability to carry a tune?" (Zhou 2005). The arguments against elections of rampant nepotism and factional infighting simply follow a certain kind of logic: since elections lack a self-evident and objectively justified standard against which to judge people and determine the result, they are unprincipled and chaotic in the eyes of the Chinese government. Hence, this fear of chaos and social disorder is a regulatory force that conditions and structures modes of official knowledge, bodies, and attitudes about the understanding of elections in Chinese society, that is, elections are not only considered as an enhancement of social involvement but also a sign of political relaxation and a way of discourse generation. With the advancement of mobile technology, elections are no longer limited to representing the will of the people but they also constitute a corrective of power through bottom-up online participation. Through this, people not only reflect their will and preferences but also conditions and constitute the other directions of sociopolitical moves. As Fortunati (2009) maintains, "Online social participation is in itself a political fact, in the sense that it shows the will of internet users to appropriate this tool for democratic participation and to build the structures and the communicative practices maybe for a future, political discussion." From this point of view, the prohibition of mobile voting can be simply understood as a preventive measure to eliminate a potential threat to governance. To some extent, we may say that the negotiation between the state and the society in digital China is no longer based on pure exploitation but on contested exclusion. The symbolic meaning of the governmental action is that the Chinese government seemingly takes a strong stand, emphasizing the myth that a new media democracy is "given" in Chinese society, in the sense that the Chinese authorities are able to define and restrain social media as principled machines with exemplary functions in order to involve people in controlled processes of participation in the interests of improving governance and enhancing regime legitimacy, that is, if anyone who might do potential harm to the process of controlled modeling, the usual practice is that he/she would simply be excluded from the information flow of the sociopolitical system.

5 Concluding Remarks

The combination of television and mobile technology provides a new experience of tele-participation, in which TV viewers are no longer passive message receivers but an active social audience who are able to participate in and influence processes of program production and distribution, in terms of the distribution of power and control shifts as well as the change of roles. In particular, the establishment of a mobile voting system in entertainment opens a space for Chinese people to express private emotions and preferences and share opinions. How the ordinary people understand and interpret this new expression channel is another issue, but it does contribute to make the show itself a social event and also enhances social participation. This emerging bottom-up and grassroots "e-sociality" not only has attracted the attention of the Chinese government but also forces the authorities to face up to and respond to the changes. On the one hand, the prohibition of mobile voting in entertainment shows is an obvious preventive measure that constrains the use of the mobile phone as merely an individual communication device. On the other hand, this policy represents a traditional Chinese view on governance and social management, in which the government does not necessarily hold an opposite attitude against "elections" and "technology," rather doing so only if the "elections" are well prepared and principled. This has eventually become an emerging and intensive space for the negotiations between the state and the society in China.

References

Bakken, B. (2000). *The exemplary society: Human improvement, social control, and the dangers of modernity in China*. Oxford/New York: Oxford University Press.

Cheng, C. T. (2009). New media and event: A case study on the power of the Internet. *Knowledge, Technology and Policy, 22*(2), 145–153.

China View. (2009). China's version of 'American Idol' gets green light. http://news.xinhuanet.com/english/2009-04/29/content_11276508.htm. Accessed 18 Apr 2011.

China.cn (2005). The fan phenomenon – Is it out of control? http://en.invest.china.cn/english/entertainment/226049.htm. Accessed 18 Apr 2011.

Danwei. (2005). The meaning(lessness) of 'Super Voice Girl'. From http://www.danwei.org/breaking_news/the_meaninglessness_of_super_v.php. Accessed 18 Apr 2011.

Fei, H. (1968). *China's gentry*. Chicago/London: The University of Chicago Press.

Fei, X. (1992). *From the soil: The foundations of Chinese society*. Berkeley: University of California Press.

Fortunati, L. (2009). *E-participation and the new media, in Cost Action 298 conference*. Paper presented at the international conference on the good, the bad and the challenging. Copenhagen, Denmark.

Goldkorn, J. (2005). The final week of TV sensation Super Voice Girls. Danwei.org. http://www.danwei.org/media_and_advertising/the_final_week_of_tv_sensation.php. Accessed 18 Apr 2011.

Hsu, F. L. K. (1981). *Americans and Chinese: Passage to differences*. Honolulu: University of Hawaii Press.

Huaxia. (2006). Super Girls' voting is another form of commercial sales. http://big5.huaxia.com/yl/kwlp/2006/00506217.html. Accessed 18 Apr 2011.

Jin, R. (1989). Yi 'xuan' dai 'ping' bu kequ ['Evaluations' with 'elections' is not to be recommended], *Renshi yu rencai*, No.3, 1989, p.25. B. Bakken (2000, Trans.). Principled and unprincipled democracy: The Chinese approach to evaluation and election. In A. Hans & T. Ngo (Eds.), *The cultural construction of politics in Asia* (pp. 107–130). Surrey: Curzon Press.

Kinnvall, C. (1995). *Cultural diffusion and political learning: The democratization of China*. Lund: Lund Political Studies.

Kuhn, A. (1992). Women's genres. In J. Caughie & A. Kuhn (Eds.), *The sexual subject: A screen reader in sexuality* (pp. 301–311). New York: Routledge.

Liu, F. Y. (2005). *Chaonu de weiyandayi* [Sublime words with deep meaning from Super Girls]. Sohu. http://club.campus.sohu.com/r-144243195-50451-0-0-10.html. Accessed 18 Apr 2011.

Martinsen, J. (2005). CPPCC: Exterminate the Super Girls. Danwei. org. http://www.danwei.org/trends_and_buzz/cppcc_exterminate_the_super_girls.php. Accessed 18 Apr 2011.

Metzger, T. A. (1998). *The Western concept of the civil society in the context of Chinese history*. Stanford: Stanford University.

Pertierra, R. (2005). Mobile phones, identity and discursive intimacy. *Human Technology: An Interdisciplinary Journal on Humans in ICT Environments, 1*(1), 23–44.

Qing, M. (2005). Super voice girls' challenges China's TV culture. *China Daily*. http://www.chinadaily.com.cn/english/doc/2005-08/12/content_468543.htm. Accessed 18 Apr 2011.

Rosenberg, S. (2006). Human nature, communication, and culture: Rethinking democratic deliberation in China and the West. In J. L. Ethan & B. He (Eds.), *The search for deliberative democracy in China*. New York: Palgrave Macmillan.

Ross, S. M. (2008). *Beyond the box: Television and the Internet*. Malden: Blackwell Publication.

Shanghaiist. (2009). Super girls coming back to Chinese TV. http://shanghaiist.com/2009/04/29/super_girls_coming_back_to_chinese.php. Accessed 18 Apr 2011.

Sohu. (2005). Sublime words with deep meaning from Super Girls. http://club.campus.sohu.com/r-144243195-50451-0-0-10.html.

Xu, J. (2005). Poking a hole in a myth of Super Girls democracy. *Boxun*. http://www.peacehall.com/news/gb/pubvp/2005/08/200508290719.shtml. Accessed 18 Apr 2011.

Yardley, J. (2005). An unlikely pop icon worries China. *The New York Times*. http://www.nytimes.com/2005/09/04/world/asia/04iht-super.html. Accessed 18 Apr 2011.

Ye, J. (2009). Chinese singing show to go on, with a new name and conditions. *The Wall Street Journal*. http://blogs.wsj.com/chinajournal/2009/05/19/chinese-singing-show-to-go-on-with-a-new-name-and-conditions/. Accessed 18 Apr 2011.

Zhou, R. (2005). Secret behind idol-making Super Girls contest. *China Daily*. http://www2.chinadaily.com.cn/english/doc/2005-08/27/content_472681.htm. Accessed 18 Apr 2011.

Zjuzhanglei. (2005). *'Gong min she hui he min zhu jing shen de xing qi – chao nu xian xiang de zheng zhi xue jie du'* [The growth of civil society and the spirit of democracy: A political interpretation of Super Girls' phenomenon]. http://zjuzhanglei.bokee.com/4020771.html. Accessed 18 Apr 2011.

Chapter 13
The Use of ICT Products and "White-Collarization" of White-Collar Workers: An Everyday-Life Perspective

Shanhua Yang and Jing Li

1 Introduction

As high-tech products, ICTs[1] (information and communication technologies) have become indispensable for sharing information and communicating interpersonally; at the same time, they have infiltrated all aspects of our everyday life, including working and leisure time, thanks to their speed, precision, meaningfulness, diversification, convenience, and effectiveness. However, as a symbol of economic globalization, ICT products have always been connected with, or aimed at, business competition and profits. As products, their value depends on consumption by consumers; as a result, producers of ICT products (corporations) have to consider how to gain as much market share as possible and how to meet consumers' demands. Thus, in some sense, the ICT products market is becoming stratified. Such a circumstance is our precondition in our study on the effect of ICT products on social life. This paper aims to discuss how the application of ICT products influences social structure, which still requires further discussion.

As a certain level of knowledge is required to use ICT products, white-collar workers, high school students, and college students have become the main consumers of these products since they were launched in the late 1990s in China. By accepting and using such products and services, white-collar workers have obtained new and different meanings in their life courses. Now that ICT products and services have so much influence on white-collar workers and their future supply—college students—the sociological significance of how ICT products and services affect

[1] Here ICT products include information technology and communication technology, all items tied to these high-tech electronic products, and the corresponding software, such as computers, the Internet, television, MP3 players, iPods, cell phones, tape recorders, digital cameras, all kinds of games, VCDs, DVDs, game software, computer software, and network software.

S. Yang (✉) • J. Li
Department of Sociology, Peking University, Beijing, China
e-mail: yangshh@ppku.edu.cn

P.-l. Law (ed.), *New Connectivities in China: Virtual, Actual and Local Interactions*,
DOI 10.1007/978-94-007-3910-9_13,

the life and attitude of white-collar workers has become apparent. Past studies on the social stratification of China were mainly on the formation of the white-collar class or the issue of the white-collar identity (Li 2004; Zhang et al. 2004); however, a thorough discussion of the mechanism of the identification process of class membership and the formation mechanism such as the influence of ICT products on membership identification is lacking. Our research attempts to fill the lacuna.

2 Concepts

Before discussing the mechanism of the identification process, a few words about the concepts on white-collar workers, new middle class, and everyday-life perspective may be useful for the analysis.

1. White-collar workers and new middle class. Past studies on the middle class, for example, by Zhang et al. (2004), term the current middle class of Chinese society, which mainly consists of white-collar workers, the "new middle class" as opposed to the "old middle class." In social classification study, the "old middle class" usually indicates the petite bourgeoisie and the self-employed, such as smallholders, shopkeepers, and independent proprietors during the early Industrial Revolution. They managed their businesses, which were based on their properties, independently (Gilbert and Kahl 1992). Although different schools vary in the definition of middle class, researchers agree on the following common characteristics of the "new middle class," namely, they (1) are generally not engaged in manual work; (2) have higher degrees and have been trained professionally; (3) live on salaries and stipends; (4) have direct control of the working process and have opinions on and influential power over public business to varying extents; (5) have a conservative political attitude, seeking democracy and fairness, and can manage the social ideology to some extent; (6) place high value on good taste, taking the lifestyle and behavior of the upper class as an example; (7) value self-achievement and self-accomplishment; and (8) possess a personal wealth, leisure time, and the ability to consume in the pursuit of a good life (such as owning a personal automobile and property, going on vacations with family members, etc.). Zhang argued that they do not occupy any operative resources but depend on "organizational capital" (managing resources) and "cultural capital" (professional skills) to obtain their social standing. Li Qiang, who similarly characterized this group, pointed out that, based on his own study, a "new middle class" is appearing in some big cities in China.[2] Based on the studies of

[2] Li Qiang claims that the basic characteristics of this group are that they (a) are relatively young, (b) generally have higher education, (c) have professional knowledge, (d) have a mastery of English, and (e) own or can work competently with computers. Most of them work in foreign-funded enterprises and newly emerging industries, so they have a distinct advantage in terms of income. Their consumer behavior reveals a high-consumption trend. The new way of life has begun to form a so-called style, the so-called petite bourgeoisie and BoBo nation. "Bourgeois Bohemia" refers to a lifestyle dominated by the middle class (Li 2004).

Zhang et al. and Li, we believe that these are the characteristics of the white-collar group in current Chinese society.[3]

2. Everyday-life perspective. At the empirical level, the temporal dimension of everyday life can be divided into general everyday life and events. The temporal dimension of events means the time consumed in handling and solving important problems which are related to survival. Such temporal processes are accidental and uncertain and may be long or short in duration. In contrast, the temporal dimension of general everyday life refers to the time that people spend living and satisfying their basic needs. The characteristic of general everyday life is its repetition and steadiness, for example, when people are cooking, eating, working, doing housework, or having a rest. Everyday life means the consumption of time; thus, either to individuals or to groups, the reality of a life course is the lapse of time (we can even say that time is the basic unit of the life course), although such lapse of time contains social meanings. In this sense, the temporal dimension of the everyday lives of individuals constitutes the form of their life courses. That is why we choose "general everyday life" as the main perspective of the study. We will use this concept to make sense of the behavior of white-collar workers and the meanings they endow to ICT products in a changing social and cultural context in China. Such meaning is mostly embodied in the formation of the mechanism of white-collar identification.

3 Analysis

3.1 The Prerequisite of ICT Products' Impact on the Chinese White-Collar Class: The Acceptance of ICT Products

There are two aspects which constitute the basis and prerequisite of the impact of ICT products on the white-collar class. On one hand, under the guidance of the "reform and open up" policy and the policy to "improve modern science and education," high-tech ICT products have received strong government support in China. On the other hand, since the implementation of the "open up" policy, China's higher education has developed fast and fostered many intellectuals who later became the solid foundation of the white-collar class. Besides, there is a close connection between the drastically reduced price and the spread of ICT products in both homes and offices, which renders them affordable to white-collar workers and even ordinary people. The distribution of netizens' educational and occupational distribution presented in Tables 13.1 and 13.2 clearly shows this.

[3] In current Chinese society, quite a few white-collar workers do not conform to the fourth and fifth characteristics.

Table 13.1 Education level of netizens[a]

BELOW high school (%)	High school (%)	Vocational qualification (%)	Bachelor's degree (%)	Master's degree (%)	Doctorate (%)
17.8	31.6	23.0	24.7	2.3	0.6

[a]According to findings published by the China Internet Network Information Center in July 2006, by the end of June 2006, there were a total of 123 million Internet users and 54.5 million computers connected to the Internet

Table 13.2 Occupational distribution of netizens

Student	Employee in an enterprise	Teacher or administrator	Government agent, party agent
36.2%	28.9%	7.4%	6.2%
Employee in an institution	Self-employed	Peasant	Unemployed
5.6%	5.3%	1.6%	8.5%
Others (army)			
0.3%			

In 1987, when virtual connectivities had just been adopted in China, there were only 300 mobile phone users in China. However, by the end of 1997, Chinese mobile phone users numbered 13.656 million, and the mobile phone penetration rate had exceeded 1%. By August 2001, the number of Chinese mobile phone users had broken through 120 million, overtaking the number in America and making China the largest mobile phone consumer in the world. By September 2004, China's mobile phone users numbered 320 million, and by early 2010 the number of users had reached 786.5 million. A recent report issued by the China Internet Network Center in July 2010 showed that the number of Internet users had reached 420 million. Behind this ever increasing number are the development of mobile communication networks and the reduction of network access fees, communication fees, and the price of handsets.[4] And, in our research, we found that for white-collar workers, the Internet plays a more fundamental role in everyday life, with the main uses of mobile phones being limited to conversation and SMS.

3.2 *Stratified Market of ICT Products*

Enterprises producing, selling, and providing ICT services generally take white-collar workers as their target consumers because white-collar workers possess high cultural taste and strong purchasing power, which are prerequisites for using ICT products. Some enterprises even claim that their products have been created for

[4]It should be pointed out that the popularity of mobile phones is also connected to the popularity of prepaid card system. Please refer to Jing Wang's article (Wang 2005).

white-collar workers. For example, MSN China provided five channels—reading, making friends, data, community, and automobiles—on the first anniversary of its establishment and thus further interpreted its brand target as the "fashionable life, white-collar gateway."[5] Such a campaign revealed that its aim was to lead the white-collar lifestyle. Obviously, companies like MSN China take the needs and character of white-collar workers into consideration when marketing their Internet services, which in turn makes white-collar workers adopt MSN as their first choice.

The white-collar choice of MSN is made through a combination of both active and passive processes. White-collar workers are more inclined to use MSN during office hours than QQ (a kind of instant message service which can be accessed either on the Internet or through the mobile phone). An informant who is a legal assistant in a Guangzhou law firm told us:

> According to my understanding, QQ is for entertainment. But MSN, many office workers like to use it. And QQ is for junior level; MSN is the kind university-level people prefer to use, such as the kind of instant messenger.

A clerk from a logistics company in Shanghai said that:

> QQ friends are friends outside of the workplace, and other companies are using MSN, not QQ.

The white-collar workers in our interviews echoed these responses: their work colleagues also use MSN. However, the interesting thing is that QQ has not been withdrawn from the market. QQ services clearly define non-white-collar workers, a more numerous group than white-collars, as their target consumers. Therefore, we can see that the formation of a stratified market for ICT products is the consequence of active consumer choice and passive acceptance led by the work environment.

3.3 Creating White-Collar Workers

3.3.1 White-Collarizing Is the Process of Creating White-Collar Workers

The reason for including this notion is that in our research we found that white-collar occupations do not equate to a white-collar group. People we interviewed felt that although they had jobs that could be called white-collar, they would not automatically be regarded as white-collar workers. A Guangzhou sales assistant said that her job and monthly income meant she would probably be considered a typical white-collar worker, but she herself thought of herself as only "junior white-collar." She believed it would take her years to really become a white-collar worker.

The white-collar class is distinguished by unique characteristics which are shown through values, lifestyle, behaviors, and even appearance. To facilitate the analysis, we combine the internal character with external manifestations and term this

[5] See articles on Jishi Net on 16 June 2006.

"white-collar understanding." Although the white-collar workers we interviewed could not say what characteristics exemplify the standards of this class, they did agree that "white-collar understanding" exists. As the aforementioned Guangzhou sales assistant said,

> The magazines that I subscribe to are mainly higher level ones and they are all for the white-collar, for instance, the *Weekly* and *Ruili*; they are very famous and their contents cover fashion apparel, cosmetics, watches, and they also have conference news, etc. …. I have read these magazines since I was young as I understand that they are for the white-collar.

With this vague but real "understanding," we are able to define the concept of "white-collarization." This refers to the process of setting out in a white-collar job to actively or passively change oneself under the guidance of the understanding to truly become a member of the white-collar class.

The white-collarization process is promoted through informal rules of the white-collar class. White-collar product advertisements are ubiquitous, with items such as cosmetics, apparel, and IT products widely promoted on train, bus, and roadside billboards; on the Internet, descriptions such as "white-collar small game" and "white-collar fashion items" can be seen everywhere. When a non-white-collar worker sees these, he or she may think, "Oh, these are the things that white-collar workers use," which in turn can place external pressure on the white-collar workers. This pressure will constrain how the white-collar workers deal with the external environment and determine their types of consumption.[6] Within the group, white-collar workers' interplay during work and leisure time and the common interests or preferences they share exert a subtle influence on their behaviors. For example, they may change from wearing very casual clothing in college to a workday formal suit, which indicates the professional requirements of an office worker. From being unaccustomed to, to accustomed to, to finally accepting formal suits, the white-collar understanding has been gradually internalized.

Whether a "quasi white-collar" or white-collar worker, one has to obey the informal rules of the group. From this point, we can say that white-collar creation is the combining process of building and being built. However, this is not to say that one necessarily has a white-collar job; if a person mimics white-collar behaviors, he or she might be called a white-collar worker, for both building and being built need a process of internalizing and shaping. ICT products play a very special role in such a process.

3.3.2 The Two-Way Interaction Between ICT Products and White-Collar Workers

When ICT products enter white-collar life, they have an impact on white-collar workers' values and way of life. And it is a two-way interaction between ICT

---

[6] The design of such products meets the needs of white-collar workers. Therefore, as with the stratified market in ICT products, a lot of white-collar workers prefer these products because they are fashionable and easy to use.

products and white-collar workers. ICT products enter the white-collar workers' lives and white-collar workers choose ICT products.

First, we will discuss the interaction between ICT products and white-collar workers: ICT products penetrate white-collar workers' everyday lives[7] as tools for work, study, and entertainment.

Indeed, as students, white-collar workers used ICT products.[8] Our respondents told us that when they were in college, they used short messaging (SMS),[9] Internet searches, online chatting, and computer games in the dorm. They used ICT products when they were young, and they knew how to use them to satisfy their own purposes and meet their own needs. Clearly, this is a main reason why ICT products become a part of their lives.

We asked our respondents how much time they spent using ICT products in their everyday lives. Many respondents claimed that more than 50% of their time was taken up by such activities. According to our interviews, the first thing many white-collar workers do after switching on the computer is to browse the news on *Sohu* or *Sina* (both very popular websites in China). The dependence on the Internet is first reflected in the dependence on information and resources. A respondent told us that 80% of the information he sources, often for work, is from the Internet. A legal assistant in Guangzhou also told us that he could not work without the Internet. He said that it was frustrating to be on leave at home without a computer linking him to the Internet; he found that he did not know anything about the world.

Second, we found that the interaction between white-collar workers and ICT products is a selective interaction. In other words, it is true that to some extent white-collar workers have a dependence on ICT products, but on the other hand, white-collar workers have cultivated a high taste since leaving college, which enables them to keep up with fashion and make their own value judgments. As a result, they can make independent choices regarding ICT products. Below are two examples to illustrate their autonomous choices.

First, they can choose useful websites and make full use of them. Internet websites are important to white-collar workers. Alumni, college, and some popular forums are most often frequented. Our respondents told us that when they graduated from college, their concerns changed and the forums they frequented were different. One respondent said:

> When we graduated and went to work, our concerns were different from when we were in college. We wanted to know more about society. And we also need to have information for our jobs. Sometimes when the information is central to our work, we do not mind registering with real names.

---

[7] We mainly research the two parts of white-collar workers' everyday lives: working (learning) and leisure time.

[8] Products like tape recorders, televisions, DVDs, and other ICT products entered white-collar life 20 years ago. However, their widespread and profound influence started from white-collars' acceptance of mobile phones which began in the late 1990s; during that period, most white-collar workers were still students.

[9] For studies on the use of SMS during SARS 2003, please refer to Haiqing Yu (2004).

Second, they develop and utilize an Internet network of "professional groups" and extend these networks to their face-to-face life. There are a lot of instant messaging groups, and most are created by professionals. A Guangzhou law firm legal assistant joined a group developed by legal workers. He said, "On the one hand, I was able to meet more people in the legal profession, and on the other, I was able to learn more about the legal industry's recent developments." Members of a specific Internet forum meet face-to-face. A senior auditor working in a foreign company told us that she had joined a face-to-face meeting organized by a forum specifically designed for the exchange of information on studying abroad. She told us that she would only join those face-to-face activities organized by the Internet forum if the meetings were useful to her. White-collar workers sometimes form a purchase group. An employee of a Shanghai car company told us that one of his colleagues gathered several people from a vehicle forum and formed a purchase group to negotiate with a vehicle company. These "professional groups" and the forum are notable in that interaction does not only occur in the network; they also develop into groups in the real world.

Finally, inheriting the advantages of traditional ICT products,[10] current ICT products incorporate "instant" and "noninstant" features (this means that if a person misses a television program, the Internet can retain the information so that he or she can view it later). This provides convenience for white-collar workers; at the same time, it makes the interaction between white-collar workers and ICT products more tool-oriented. For example, some white-collar workers watched a very popular TV show called *Super Girl*, but they did not enjoy watching the whole show. They just wanted to know who won the show in order to gain material for communication in their social lives.[11] They focused on results rather than on the process. As a result, through the interaction between white-collar workers and ICT products, it was clear that ICTs are functional for the white-collar workers. And the white-collar workers are not easily dominated by the ICT products as they can make their own choices in using them. However, ICT products and the relevant services also provide information resources and the channels through which they can be acquired. The most up-to-date information is central to the white-collar workers, particularly for their work. Thus, white-collar workers are, in this sense, actively pursuing the latest ICT products. This point is the second meaning of the two-way interaction. As one of our white-collar respondents maintained, "As for the latest products, I do not necessarily pursue them, but I have to understand them. It is very important in this information age!" He continued, "What you read determines your identity; what you wear determines your status!" This respondent has pointed out precisely that the pursuit of fashion and up-to-date information has constructed white-collar workers' identity

[10] We regard radio, broadcasting, and television as traditional ICT products. They have a common characteristic: they cannot be repeated, so the viewer has to watch or listen to a program as it is being broadcast.

[11] An interviewee said, "I definitely won't vote. It is total nonsense" (GBN, Assistant Lawyer, Guangzhou).

and status, and hence the boundary between the white-collar class and other social classes can then become clearer.

3.3.3 Internal Identifying Mechanism and Its Formation Within the White-Collar Group

As mentioned in the previous paragraph, after realizing the existence of a "white-collar understanding," white-collar workers gradually and persistently bring their behavior much closer to the understanding. This may be due to group pressure or active choice. The interaction between white-collar workers and ICT products accelerates the pursuit of fashion by white-collar workers, presenting their consumption characteristic. Consumption ability and consumption options represent a group's inner character best because they endow the meaning of status. Therefore, the interaction constitutes the basis of an identifying mechanism.

We believe that white-collar workers' strong consumption ability is a prerequisite for the identifying mechanism. White-collar income is relatively high, enabling them to purchase the products and services that match their careers and identities. Taking ICT products as an example, when white-collar workers choose ICT products, they focus more on appearance and functions than price. In 2001, an informant who worked in a media company in Beijing told us he had five mobile phones, and this was totally within his financial ability. He focused on the musical function and the appearance of the mobile phones. He said: "Most of the MP4 devices on the market can only save one movie, and it is very troublesome to copy to it, and for those cheap ones, the screen is too small." This shows that white-collar workers' expectations are usually very high. They tend to think, "I am a white-collar worker, so I should buy products that match this class level, so I will match the characteristics of the white-collar class more."

The formation of an identifying mechanism is the process of combining one's personal particularity with group commonality. We know that the white-collar group members are people who pursue their group identity, and they are capable of showing their unique character. But this pursuit would easily be constrained in their workplace. A staff member working in a Beijing foreign enterprise always complained that she had no personal clothes as she had to work wearing clothes of a style specified by the company. So white-collar workers can only show their personalities within the group commonality. For instance, white-collar workers prefer phone calls to SMS as their method of communication because they are easier, more direct, and can get the job done more conveniently; this practice has changed their personal ways of using mobile phones. However, another example shows the particularity in the commonality. Another informant who is a professor in a university in Guangzhou maintained, "I use the 163 email and for normal usage it is free of charge, but I have paid to upgrade it to a VIP account. It is very useful as if you have a VIP account, you can receive email on your mobile phone." This professor uses email as other professors do. Using email is a common practice. But he prefers a faster service and pays for "email on the go." Particularity is in commonality.

4 Brief Summary

Based on the above analysis, we can draw the following conclusions:

1. One direct consequence of the mechanism of white-collar workers' inner identification is that the boundary of white-collar groups is becoming clear. Although the emergence of a "white-collar class" or "new middle class" is still being debated, as is the question of its members' social strength, we believe that, after the process of "white-collarizing," the group of white-collar workers will gradually stabilize or solidify. If others hope to join the group, they have to experience the process of building and shaping themselves. Such a consequence cannot be separated from the interaction between ICT products and white-collar workers. In some sense, we can even find that it is the interaction between ICT products and white-collar workers that solidifies the white-collar class.
2. From the definition of the "new middle class" (almost equal to white-collar workers), we find that the past study of social stratification usually chooses and fixes the indexes of the classification through outer characteristics of groups (occupation, salary, social standing, lifestyle, and behavior); however, we believe that the establishment of a social class requires a kind of inner identification and exclusion to clarify the group boundary. While some past studies mentioned this point (Zhou and Tian 2001), how a social class forms such a mechanism and what the meaning of this mechanism is have been ignored. The influence of ICT products on the "white-collarizing" of white-collar groups offers us a perspective on such questions and provides us with a route to discuss the influence of ICT products on other social groups (such as blue-collar workers). Our study in the field of social classification is therefore significant.
3. When we define the concept of "everyday life," we assume that daily life means the consumption of time; thus, either to individuals or to groups, the reality of a life course is the lapse of time (we can even say that time is the basic unit of a life course), although such lapse of time contains social meanings. In this sense, the everyday lives of individuals constitute the form of their life courses. Thus, when ICT products fully intrude in the everyday lives of white-collar workers, their life courses inevitably show characteristics from other social groups. The main hypothesis of life course theory is that social change influences a cohort, although social structure also restricts the life course of a cohort. For example, although both white-collar workers and migrant workers who were born in the 1980s have experienced social transformation since the 1990s, they have fundamentally different life courses. In some sense, such a track is a symbol of social stratification. As a result, this would probably enlighten us in understanding the theory of life course in another new dimension.

References

China Internet Network Information Center. (2006). The 18th statistical report on internet development in China. http://www.cnnic.net.cn/. Accessed July 2006 (in Chinese).

Gilbert, D., & Kahl, J. A. (1992). *The American class structure*. (H. Peng, Trans.). Beijing: China Social Sciences Publishing House (in Chinese).

Jishiwan. (2006). MSN anniversary: Pursuing fashionable life, creating white-collar gateway. http://cn.msn.com. Accessed 16 June 2006.
Li, Q. (2004). The new trend of social classification's structural change in current China. *Jiangsu Social Sciences, 2004*(6), 93–99 (in Chinese).
Wang, J. (2005, October 20–21). *Development of mobile phones in China*. Paper presented at the international conference Mobile Communication and Asian Modernities II, France Telecom, R & D, Beijing.
Yu, H. (2004). The power of thumbs: The politics of SMS in urban China. *Graduate Journal of Asia-Pacific Studies, 2*(2), 30–43.
Zhang, W., Li, W., & Gao, G. (2004). Structural characteristics of new middle class in current China. *Jiangsu Social Sciences, 2004*(6), 100–107 (in Chinese).
Zhou, H., & Tian, G. (2001). Worker groups' stratification and identifying differences (unpublished). "Tiaozhan Cup" paper, Department of Sociology, Peking University.

Chapter 14
Home and Away: A Case Study of Students and Social Media in Shanghai

Larissa Hjorth and Michael Arnold

1 Introduction

"If I was a fish then QQ would be my water"
—25-year-old, male postgraduate student

In 2009, Internet penetration rates in China reached 298 million users and surpassed the global average level for the first time (CNNIC 2009). Over the last 2 years, CNNIC statistics have noted a sharp increase as lower-income and less-educated people are also becoming Internet-connected. As the Internet reaches the less-privileged in addition to the more-privileged, and reaches people outside of the major cities, motivations for using the Internet and access routes to it diverge. For many less-privileged and for those outside of the big cities, the only access route to the Internet is not via the computer but via the mobile phone. These alternative routes to the Internet are catered for by China's SNS, such as the oldest, QQ, which caters to diverse platforms and modes of access to its numerous services—in particular, affording a specific version of mobile Internet.

While technological infrastructure and access to the Internet are still very much an urban phenomenon, changes have occurred in rural areas with the number of rural Internet users reaching nearly 85 million (CNNIC 2009). In short, while the number of urban Internet users exceeds the number of rural Internet users, the growth rate of rural users now far exceeds that of urban users. While the Internet is mainly accessed by the middle and upper class in urban city areas via personal computer, the rest of China's predominantly working class demography deploys

L. Hjorth (✉)
School of Media & Communication, RMIT University, Melbourne, Australia
e-mail: larissa.hjorth@rmit.edu.au

M. Arnold
School of Historical and Philosophical Studies, University of Melbourne,
Melbourne, Australia

P.-l. Law (ed.), *New Connectivities in China: Virtual, Actual and Local Interactions*,
DOI 10.1007/978-94-007-3910-9_14, © Springer Science+Business Media B.V. 2012

GPRS (General Packet Radio Service) for Internet access via the mobile phone. In China's undulating twenty-first-century social and economic landscape, class (as a socioeconomic category) continues to be a dominant factor influencing the types of emergent technocultures. Class shapes, and is shaped by, the differing technocultures constituting China (Qiu 2008; Donald et al. 2002). Through these shape-changing technocultures, we can see numerous forms of mobility that complicate notions of age, class, place, and identity.

In this rise of Internet access accompanying economic and geographic movements, the increased uptake of social media is notable, and in this changing technoscape, numerous types of SNS—each interpolating different classes, communities, and lifestyle clusters—can be observed. There are many differences among Chinese SNS, but the differences predominantly fall into two technocultural trajectories—sites such as Xiaonei/Renren (akin to Facebook), Kaixin, and MSN are accessed via computer, are often simultaneously opened and used on someone's desktop, and are typically used by female "white-collar" workers in the cities. On the other hand, Fetion and the aforementioned QQ can be accessed both via mobile phones and computers and attract a broader socioeconomic demographic.

In particular, QQ occupies a special role in the technocultural imaginary of China. QQ is not only one of the oldest SNS in China—and thus, for many, their first SNS and their introduction to social media—it is also an SNS predominantly deployed by less-privileged socioeconomic users and by rural and small town users. QQ's various applications, in particular "chat," allow families to keep in contact regularly for little or no charge. These convenient applications, broad user base, and low costs are but three of the many reasons why QQ has attracted over 376,000,000 registered accounts (www.web2asia.com). Another important reason can be heard in the voices of users who regard QQ as being Chinese *before* it is technological (Koch et al. 2009). That is, QQ is clearly identified as a national product and represents China's widespread nationalist sentiments in the context of the Internet, something no other SNS can do.

Moreover, QQ is emblematic of the new breed of young media literates, known as the *ba ling hou* (born between 1980 and 1989), who are migrating geographically from the regions to the cities, are migrating socioeconomically through educational and IT policies (CNNIC 2009) and, as they do so, are transferring their media knowledge to their parents and grandparents in order to maintain kinship ties. Unlike the other SNS (Xiaonei/Renren and Kaixin) that are mainly for and by university students, QQ's lineage and attendant demography gives it a specific role within the lives of the young. QQ's particular material characteristics (it is free, convenient, has a broad user base, and can be accessed through the phone), together with its symbolic characteristics (it conjures feelings of nationalism and nostalgia as China's oldest SNS), provide a pathway for these young media literates to not only keep in contact with family while studying away from home but also afford older generations of users a discursive, nonofficial route into new media literacy. Often, parents and grandparents begin using QQ to keep in contact with their child studying away from home and, before long, develop broader skills with online technologies, including participating in gaming and informational services.

Drawing from a preliminary study of university students, now studying in Shanghai but who derive from elsewhere in China, we explore some of the ways in which this new generation of media literates are using SNS to keep in contact with family and friends at home and, as a consequence, are transferring their new media knowledge to their parents and grandparents, many of whom have not used a phone, let alone the Internet, before. Though discursive routes such as mobile phone Internet via GPRS, these older generations—often illiterate, from lower-socio-economic backgrounds, and living in country towns—are able to gain entrance into twenty-first-century technocultural practice.

We argue that QQ, in particular, has become a glue for cross-generational mobility, marking a new pathway of lifestyle cultures in China. Just as migrant workers—or what have been called the "floating population"—use new mobile technologies to reinforce social networks and kin relationships in practices called "communities of the air" (Wei and Qian 2009), the usage of QQ by university students is marking new forms of cross-generational media literacy that is helping bridge the gap as they migrate, via education and IT policies, into a new lifestyle.[1] While these university students from lower socioeconomic backgrounds coming to Shanghai to study cannot be defined as "working class" (Donald et al. 2002), they do represent a small component in China's changing social fabric (Dong 2003) in which labor, capital, and technologies are being mobilized—with both beneficial and detrimental outcomes (Hau and Salili 1996).[2]

This chapter is a preliminary study exploring these new forms of mobility in which young adults from the *ba ling hou* generation impart their media knowledge to their parents and grandparents. In this study, we specifically focus upon the students' discussion of the media usage. A future study will involve interviewing parents and connecting their new media literacy and experience with that of their "teachers," the children.

---

[1] While Shanghainese middle class enjoy the luxuries of Internet-enabled computers at home and on the move in addition to mobile devices that often have 3G capacities, for the working class migrants, their mobile phone is THE access to both their new working life in Shanghai and their family and friends back home. For many working class and students, 3G mobile Internet, despite the hype (CNNIC 2009), is not a reality. Rather, through the deployment of various methods including GPRS, WiFi, and PCs, these users are demonstrating new forms of "making do" practices indicative of the growing demographies of "have-less" (Qiu 2008). These making do practices of the have-less are amplified when exploring Shanghai's undulating and ever-changing technocultures and the attendant forms of mobility.

[2] For those of the working class, moving from small towns to large cities for work has become a reality in which mobile technologies feature predominantly as a way to maintain ties with family and friends back home as well as helping with new work-related connections (Wei and Qian 2009). As noted by Jack Qiu (2008), China's rapidly changing social informationalized processes that have seen technologies such as the Internet migrate from the "elite-monopolized" stage—that is, the preoccupation of the middle class—to the new stage characterized by the deployment and reappropriation of media technology by the working class. In this changing social fabric in which various forms of mobility (geographic, cultural, socioeconomic) can be witnessed, the significance of new media such as QQ in helping ensure kinship ties cannot be underestimated. Far from the new media distancing them from their past and socioeconomic background, it provides a material and symbolic portal between home and away.

This study highlights that as social media evolves, it is far from a western or youth prerogative as suggested by the focus of SNS literature to date.

Moreover, the study highlights the need to understand *Online China* inclusive of the politics of the personal, rather than just focusing upon the Internet as a site for the expression of public opinion and official party politics. *Online China* needs to be understood in terms of the unique technocultures that constitute the Chinese experience, but have also been pivotal within the global chains of ICT (Information and Communication Technologies) production and consumption over the last couple of decades (Robison and Goodman 1996; Donald et al. 2002). In commercial terms, China's role globally in the production and consumption of ICTs has focused on its role as a hardware manufacturer, a media pirate, and a growth market. In social terms, much of the literature on China's internal Internet deployment has focused on its role as a contested site between public opinion and government policies (Damm and Thomas 2006; Hughes 2003 Tai 2006). Like South Korea, the Internet in China has provided a way to engage with, observe and interpret struggles for political change, with social media such as blogs attracting much focus (Bruns and Jacobs 2006; Lovink 2007; Yu 2007; Dong 2003), while seemingly "less" political media such as SNS have been relatively overlooked (Koch et al. 2009; Yu 2007), despite often being deemed threatening and thus banned by the government (for example, in 2009 Facebook and Twitter were barred).

China's technoculture deploys numerous examples of SNS, both local and global. Local SNS include QQ, Kaixin, Renren (previously Xiaonei), and Fetion; while MySpace and MSN feature as global examples. However, the use of both local and global SNS is only apparent in "cosmopolitan" cities such as Shanghai. For the general population of China, few SNS are used or even known apart from QQ. Part of QQ's success has been ensured by its inclusion and convergence of various platforms from chatting and SNS (QQzone), to online, localized news. Via Internet portals its service is free, and its various platforms such as QQ chat can be adapted to devices such as mobile phones. This is important given that while most of the population outside of the big towns and cities do not have the Internet they all have a mobile phone. QQ represents the largest cluster of SNS services in China, resulting in QQ being among the largest in the world. However, despite the significance of QQ in emphasizing the vernacular and local of Chinese sociality—and its history as the first SNS and IM that ushered a new generation of digital natives through to Web 2.0—studies of QQ have been limited.

As Koch et al. (2009) note in their study "Beauty is in the Eye of the QQ User: Instant Messaging in China," "the emergence of this local homegrown QQ software indicates that it plays an important role in Chinese society" (2009: 265). It is for this reason that QQ can help to "understand the influence of this specific Internet platform as situated within Chinese culture" (2009: 265). As they note, the various services and platforms of QQ mean that it differs greatly from the "Internet" in China in general. Rather, it is the portal between media, between platforms (mobile and Internet), between generations, between media experiences, between cities and the regions, and between classes.

Thus, through the lens of such SNS as QQ, we can chart this transitory period in which China's technocultures give rise to emergent forms of mobility—especially around the new breed of youth that have accompanied this rise. Through QQ, one can examine a variety of mobilities, but in this case study, we will focus upon the current generation of university students, studying in Shanghai, that come from lower socioeconomic backgrounds. We explore how QQ helps to maintain the bond with their family and friends back home while also being the lynchpin between their working class backgrounds and their upwardly mobile future careers and lifestyles. Through the coordination of various social media networks, they are able to both enhance existing relations and also develop new ones with fellow students and part-time work colleagues. In this ensemble of various SNS and IM, it is QQ that remains the enduring network for maintaining familial and old friendships.

2 Home and Away: A Case Study of Fudan University Students

> "If we was a fish then QQ *[social media]* would be my water."
> —Bao, 25-year-old, male postgraduate student

2.1 The Ba Ling Hou Generation

In the undulating technocultures of China, Shanghai represents a particular version of Chinese media practice (Wang and Lau 2009). Over the last 10 years, cities such as Shanghai have not only seen the implementation of new technologies in educational and work settings but also an increasing trend of studying away from home. Born in the 1980s, the *ba ling hou* generation is the first to grow up with the Internet and new media as part of everyday life and is the first to appropriate this form of mediated mobility. They are also the generation that have been allowed to move from the regions to the city to study—a concept unheard of for previous generations. Often from one-child families, these *ba ling hou* are incredibly close to their parents, and although geographic mobility for education is a given for this generation, there is a need for continual contact with friends and family back home. In studying away from home, the deployment of mobile media—especially for SNS like QQ and Renren—has been integral for a generation that negotiates *home and away.*

In this section, we focus on cross-generational gaming through case studies of students who migrate, often from small towns in China, to Shanghai, to study at university. In this migration, the deployment of mobile and social media plays a pivotal role in their maintenance of sociality. As the first generation to grow up in China's emerging net culture, the *ba ling hou* are a product of China's first educational

IT initiative—CERNET (Chinese Education Research Network). This initiative constructed a national educational network, which began construction in 1994 and was taken up and continued in 2000–2010 through the EISS policies (an acronym for "Electronic Information Service System" or in Chinese "xiaoxiaotong") through which the government orchestrated the implementation of computers and online education within primary and secondarily schools. The EISS was to enable 90% of the independent middle schools and primary schools throughout the country have access to the Internet, along with the deployment of online resources to be shared among teachers and students. The average *ba ling hou* student is a member of the EISS generation, growing up with the benefits of these policies, encouraging high media literacy and Internet use as part of everyday life. Current Fudan University students are a good example of these new media literates and their attendant forms of mobility and copresence.

CERNET and EISS opened up sociotechnical possibilities for the *ba ling hou*, and our informants' experiences reflected the shifting technocultural landscape and their mobility within it. For example, a female postgraduate student (aged 26) spoke of her time at Fudan University as a period marked by these shifts in technological access and availability. She noted that in her first year (2004), there were few computers and Internet portals available in classrooms; however, within a matter of 2 years, she observed a dramatic increase in both computers and accessibility—to the point now where at the end of the EISS project, everyone has a computer and Internet access is all-pervasive.

Many of the *ba ling hou* students are from working class regional backgrounds, and the socioeconomic and sociotechnical mobility they have achieved through education is being transferred to their parents and hometowns in unofficial ways, through cross-generational media literacy.

3 Fieldwork

In June of 2009, we conducted fieldwork examining *ba ling hou* students' usage of mobile and social media, and in June and July of 2010, we visited again and interviewed students and their parents, talking with each group separately. Through a variety of methods, most importantly group discussions and one-to-one interviews, we found that a majority of our students were not from Shanghai and came from a lower socioeconomic background, and many spoke at length about the deployment of mobile and social media to negotiate their experience of being home and away. Moreover, this need to maintain sometimes daily contact with their parents back home not only required the students to use a variety of mobile and social media but also to teach their parents how to use these media. A female informant, aged 19, talked of cross-generational media literacy in this way:

> They have a lot of time to stay at home, so they will play the computer games and want to surf online. But my father and mother are not good at it yet, so we continue to teach them and with the help of QQ, we can contact them more often. For example, when we come

> back home, I find that my father's mobile phone has something wrong—it always happens and he can't receive my short message. I said he's a little old for it. He has played games in QQ and also, *Happy Farm*. He liked stealing vegetables. But my uncle is more of social media user. Several years ago, I taught him how to use the Internet, how to connect—how to talk with others, by QQ or something like that. And now he uses it all the time. He even makes friends with strangers. And so, every time we come back to my home and we can talk a lot about this QQ, and games. I don't know if the technology is a very good thing for him. I don't know, because we think maybe he has spent a lot of time on this new technology, maybe too much. But from my side, we have no doubt he has a very, very young heart from this technology side. And we think he is enjoying his life very much.

For many of the respondents, the role of QQ was so integral to everyday life that it was viewed as second nature—almost akin to nature. QQ has become a "natural" backdrop for contemporary Chinese technocultures (Koch et al. 2009). As one female respondent, aged 25, noted, "I think QQ is like air. When you breathe it you feel nothing special but without it, you can't survive." For one male respondent aged 24, he echoed the previous respondent by stating, "If I was a fish then QQ would be my water." Indeed QQ seemed to function like the air and the water for the media ecologies and attendant forms of media literacy in China today. Not only is QQ in almost all cases the first SNS for the current generation of university students, the *ba ling hou*, it is the one that they continue to use to maintain kinship relations and older ties despite the uptake of other SNS. In this way, QQ can be seen as a set of emerging cultures, practices, and analogy for China's burgeoning twenty-first-century technocultures, a mediascape whereby the dominant rise in the population of users are from lower socioeconomic backgrounds.

This generation has witnessed major technocultural shifts during their university studies. As one female postgraduate student (aged 26) noted, much of her time at Fudan University as an undergraduate (she is now a Masters student) was a period marked by shifts in technological access and availability. Indeed, at the end of the EISS project, the impact is significant, especially for the current young generation of university students for whom the Internet is a pivotal part of everyday life. For many, imagining 1 day without the Internet is more than difficult and inconvenient; it is unimaginable. It is this new generation of digital literate that are helping usher older generations through the new technoculture, as well as being the teachers for the working class as they embrace the making-do mobile Internet technocultures of China.

Many of the *ba ling hou* we interviewed used the Internet to a point where they felt the need to restrict their usage to specific time brackets, in order to counteract possible Internet "addiction." While many expressed their fear of being addicted as a social faux pas, a few freely embraced their highly dependent relationship on technology by studying in disciplines such as Informational Technologies or Journalism, where use of technology was a necessary given. For those with access to a variety of SNS, clear technocultural differences were evident across the SNS landscape. For example, MSN was used for the first time by many of our informants in the context of their part-time workplace environments or last year internships. While they were not permitted to use QQ (given its "social" tool association and perhaps its class and regional connotations), employers strongly supported the use of "professional" media such as MSN. One female student, aged 22 and studying Journalism, considered herself lucky to have a job in which MSN and QQ usage was encouraged.

As another example, a female student aged 23 and majoring in Chinese studies and literature observed vast differences in the levels of trust and disclosure associated in the various forms of SNS. In particular, QQ is viewed as the SNS for copresent intimacy between close friends and family, while other SNS such as Renren (was previously Xiaonei) are viewed as more public and thus not an avenue for expressing feelings and private thoughts. For another female respondent aged 20,

> I don't use MSN because I have no need (no friends using it). I often use QQ because when I was 13 years old I would often communicate with my family such as my brothers, sisters, and parents via QQ because we chat via video and see each other. But I use Fetion to only send messages to my students because I am a tutor for the undergraduate students. I use Xiaonei only for my classmates whose relationship is not very good in the past. Because with the good friends we will often send messages via QQ and by mobile phone (SMS). On Xiaonei I don't contribute my feelings.

Given that all but one of the focus group participants was from elsewhere than Shanghai, our conversation was dominated by issues around family, home, and how the different SNS technocultures functioned to facilitate this within the various forms of mobility—class, technological, geographic, and social. For one female postgraduate student aged 24, one's origin strongly influenced what media one used. She noted changes of SNS dependent on life stages (work, etc.) and where one is living. She noted that QQ continued to play a special role in maintaining her familial and kinship ties, and in order to do so, she had taught her family to use QQ. This was a common pattern found in almost all of the participants—regardless of different SNS usage, QQ continued to function to maintain kinship ties and geosocial and economic connections with their past. One female respondent, aged 27, notes,

> I think your relationship with technological communication tools depends where you are in China and where you are from.... After I finished my Bachelor degree at Taijin University I came to Shanghai (Fudan University). I used QQ a lot when I was in university and then I started using MSN when I was in senior year because I was doing an internship. In the office, MSN is really important to transfer documents.... For me, Xiaonei is not a place to make new friends, just to reconnect to already known friends and contacts.... But I continue to use QQ for my parents. Last year they didn't know how to use a computer, let alone QQ, so I showed them. Now my family has moved to a new place so we use QQ video so we can see each other whilst chatting.

When asked to discuss how migration had affected their usage of SNS, the focus group reviewed the various SNS and discussed how they were deployed and the nature of their associated social networks. Many were in agreement about the different usages of MSN and QQ and the attendant class and lifestyle associations. Issues such as class, place, technological access, and language arose as important factors ensuring QQ's ongoing success and relevance for both young and old generations entering Web 2.0 mediascapes. Many noted that QQ offered the most options through its various localized services. For one female respondent aged 26,

> I noticed when I came to Shanghai people were using MSN and Skype which I hadn't seen used elsewhere…. In other parts of China there are a lot of people who don't have a computer in their home, let alone Internet. So QQ is much more popular with these people as it can be accessed via the phone mobile—something that everyone has.

For another female respondent aged 27, QQ's success was not only due to its easy accessibility across platforms and technologies but also due to its abilities to localize its services. Features such as the QQ news service that focused on the region that the user was accessing the service from were viewed as very important in defining a sense of place. She notes,

> When you open QQ the news pops up, for example the earthquake in Sichuan province, it is news that Chinese people are more interested in…. While the white collar may use MSN more often, she will also have a QQ account. Different geography will influence what people use. Geography influences the hardware—some places don't have computers and Internet access, they just have a mobile phone. Or they may have the hardware and not the software. In country areas computers are not very common so often if they want to use the Internet they must go to a café. Some villages only have one Internet café. Whereas every person has a mobile phone. In rural areas, they can only use the mobile phone. Also, many older generations use QQ—parents, grandparents. They use it on the computer. On the phone they feel they don't want to leave a message because it might be too troublesome.

Here, the point about not wanting to leave a message has little to do with cost (given that QQ's various services such as QQ chat are cheap, if not free, dependent upon device used), but rather, a Chinese social protocol whereby leaving a message obliges the child to call back. Gift giving has always played a significant role in the maintenance of relationships in China, and this instance highlights the fact that alongside the changes evident in the new technocultural landscape, continuity is also evident, and many gestures with ICTs mirror older patterns around obligation and reciprocality (Taylor and Harper 2002). The significant role of QQ in facilitating kinships can also be seen as helping to secure metaphoric kinship ties between partners. As one female, aged 23, notes:

> Actually I think I know nothing about the technology… but I cannot live without the technology because I use the mobile phone and Internet to keep contact with the outside world. Everyday I have to use the Internet to check my email, and check for course (university) postings. But one of the most important things for using the technology is because I miss my boyfriend… he is living in a different city so everyday I have to use QQ to have contact with him.

Here we see the role in which different SNS represent various groups and user's modes of mobility. For this above user, the physical dislocation from her partner is alleviated through the contact via QQ. QQ becomes a significant portal in the maintenance of the relationship and overcoming the geographic distance. For another respondent, female aged 22, the usage of QQ every day provided a familiar experience in an unfamiliar environment (as she moved away from home to attend university). It also helped to maintain her contact with her family, subsiding the homesick feelings:

> I used QQ almost everyday. I use both my computer and mobile phone all day, every morning and every night. I use various services of QQ. This is the first year I have lived away from home and so I use QQ mainly to keep in contact with my family at home as I board away from home. I use QQ almost everyday to make contact with my family. I get very

> homesick and so QQ helps with that. Last month my mother had an operation and I couldn't come home because of exams so instead I used QQ to keep in contact with her continuously.

In the face of these various forms of mobility (class, geography, social, and economic) in China's transforming technoculture, the pivotal role provided by QQ remained the constant. It connects generations, migrating across geographic distance, as well as securing and locating a sense of place, family, and community. To cite again one of the respondents, the significance of QQ in the shifting technoculture cannot be underestimated: "I think QQ is like air. When you breathe it you feel nothing special but without it, you can't survive."

4 Still Mobile: Conclusions on China's Mobile Technocultures

With the rise of networked, mobile technologies no longer just the preoccupation of developed countries, we are witnessing a growing diversity of instances in which place matters. In each location, what constitutes mobility (and immobility), is different. These contingencies are shaped by technological, sociocultural, linguistic, and governmental factors to name a few. Far from eroding a sense of home and kinship ties, mobile technologies are reinforcing the multiplicities of what constitute a sense of place. As social geographer Doreen Massey (1993) observes, a sense of place is more than a physical, geographic experience. Indeed mobile, networked technologies—that is, mobile media—not only transform how we experience place in everyday life, they also highlight that place is more than just a physical, geographic notion. Thus, despite the burgeoning of numerous forms of mobility—geographic, technological, socioeconomic, and physical—as part of global forces, mobile media are helping to facilitate the significance of place. That is, place as a space that is not only geographic but evokes cartographies of class and family.

As the *ba ling hou* move away from home in the quest for a university education, often there will be many feelings of homesickness. The mobility of the lower socioeconomic children through social media is a growing phenomenon given the new media mobility afforded through policies such as EISS. For the *ba ling hou* studying in another place, SNS offers these youth efficient and convenient ways in which to keep connected with family and friends at home, helping to alleviate some of the feelings of loneliness and sadness.

In particular, through QQ, university students away from home are teaching their parents how to be online in order to communicate frequently. QQ is not only the first SNS used for the young, it is now the portal that introduces older generations to SNS. Moreover, for many, this introduction to SNS is *also* an introduction to the Internet and, in this way, could be seen as an important rite of passage in beckoning a new breed of cross-generational new media literacy. For illiterate parents or grandparents, features such as QQ video and games like *Happy Farm* allowed them to see each other and talk for free. Thus, far from SNS being vehicles for endorsing their new mobility in which the past is forgotten, it provided a portal for reconnection and rerooting with one's past. While SNS like Xiaonei/Renren are predominantly used by university students to connect with new and old "nonclose" friends and networks, many of the other SNS, especially QQ, is being used to maintain older, more intimate, relationships.

This case study of university students in Shanghai is the first in many that need to be conducted if we are to understand the relationship between these emerging forms of mobility and China's twenty-first-century technoculture. As noted earlier, a follow-on study should focus upon the impressions and experiences of both the parents and children as they accompany each other through this emerging rite of passage. While Shanghai has long been a location in which working class university students from elsewhere come to study and transgress their socioeconomic background, the role of ubiquitous technology in this equation is new. As the first generation to grow up with the Internet, this generation—and their relationship to other generations—can provide some new models for media literacy. Moreover, while these new media literates are demonstrating a new form of mobilities across geographic and socioeconomic differences, they are highlighting the importance of maintaining a connection with their past. As a lens onto emerging forms of intimacy, kinship, and media literacy, SNS demonstrate that in the increase of various modes of mobility, home (or a sense) is never far away.

References

Bruns, A., & Jacobs, J. (Eds.). (2006). *Uses of blogs*. New York: Peter Lang.

CNNIC [China Internet Network Information Center]. (2009, January). Statistical survey report on the Internet development in China. http://www.cnnic.cn/uploadfiles/pdf/2009/3/23/131303.pdf. Accessed 14 June 2009.

Damm, J., & Thomas, S. (Eds.). (2006). *Chinese cyberspaces technological changes and political effects*. New York: Routledge.

Donald, S. H., Keane, M., & Hong, Y. (Eds.). (2002). *Media in China: Consumption, content and crisis*. London: Routledge.

Dong, D. (2003, May 27). Resistance within contexts: A study of university BBS youth culture in China, *International Communication Association*. Marriott Hotel, San Diego, CA.

Hau, K. T., & Salili, F. (1996). Achievement goals and causal attributions of Chinese students. In S. Lau (Ed.), *Growing up the Chinese way* (pp. 121–146). Hong Kong: The Chinese University Press.

Hughes, C. (2003). *China and the Internet: Politics of the digital leap forward (Politics in Asia Series)*. London: RoutledgeCurzon.

Koch, P., Koch, B., Huang, K., & Chen, W. (2009). Beauty is in the eye of the QQ User: Instant messaging in China. In G. Goggin & M. McLelland (Eds.), *Internationalizing Internet studies* (pp. 265–284). London: Routledge.

Lovink, Geert. (2007). Zero Comments: Blogging and Critical Internet Culture, London and New York: Routledge.

Qiu, J. (2008). Wireless working-class ICTs and the Chinese informational city' [special issue]. *The Journal of Urban Technology* on 'Mobile media and urban technology'. L. Hjorth, G. Goggin & A. Crawford (Eds.), *15*(3), 57–77.

Robison, R., & Goodman, D. S. G. (Eds.). (1996). *The new rich in Asia: Mobile phones, McDonalds and middle-class revolution*. London: Routledge.

Tai, Z. (2006). *The Internet in China: Cyberspace and civil society*. New York: Routledge.
Taylor, A., & Harper, R. (2002). Age-old practices in the "New World": A study of gift-giving between teenage mobile phone users. *Changing our world, changing ourselves* (pp. 439–446). Proceedings of the *SIGCHI* conference on Human Factors in Computing Systems, Minneapolis.
Wang, J., & Lau, S. Y. (2009). Gentrification and Shanghai's new middle-class: Another reflection on the cultural consumption thesis. *Cities, 26*(2), 57–66.
Wei, D., & Qian T. (2009, June). *The mobile hearth: A case study on new media usage and Migrant Workers' Social Relationship*. ANZCA conference, Brisbane.
Yu, H. (2007). Blogging everyday life in Chinese internet culture. *Asian Studies Review, 31*(4), 423–433.

Part V
New Connectivities and Chinese Social Context

Chapter 15
Toward a New "Electrical World": Is There a Chinese Technological Sublime?

Matteo Tarantino

1 Introduction

The massive penetration rate of information technologies in China in the last decade has left many Western commentators first astounded, then enthusiastic, and eventually puzzled. While China appears to have eagerly embraced digital communication technologies, the predicted outcomes in terms of social change (mostly, democratization and civil society development) have not yet materialized, or at least not in the terms predicted or hoped for by scholars and commentators. Chinese society appears to be making a symbolic investment in technology quite different from that operated in the West. This chapter intends to present a broad framework to investigate how Western and Chinese culture give meanings to communication technology, in order to shed some light on the apparent contradictions that are all too often explained away as products of the exercise of authoritarian control over "liberating technologies." The key concept is that of the technological sublime, that is, a form of intense collective emotional participation in the experience of technology (Marx 2000; Nye 1996), the development of which we will sketch in the first paragraph. We will then present our argument in three steps: first, we will try to investigate the philosophical foundations that gave birth to the Chinese technological sublime; we will then examine how it shaped the Chinese technological imagination through some examples taken from Chinese science-fiction literature. Finally, we will illustrate, using empirical research, how the Chinese technological sublime may contribute to the shaping of individual processes of meaning attribution toward technology.

M. Tarantino (✉)
Department of Sciences of Communications and Performing arts,
Catholic University of Milan, Milan, Italy
e-mail: matteo.tarantino@unicatt.it

P.-l. Law (ed.), *New Connectivities in China: Virtual, Actual and Local Interactions*,
DOI 10.1007/978-94-007-3910-9_15, © Springer Science+Business Media B.V. 2012

2 The Situation in the West

During the last 15 years, authors such as Stefik (1997), Mosco (2004), Dery (1997), Davis (1999), and especially Noble (1997) explored the ways in which Western culture makes sense of technology by framing it within broad cultural tensions. One of the main cultural tensions is that between man and nature, which lies at the root of Western religious thought. This dichotomy states that human beings once belonged to the divine, but that because of their individualistic nature, they eventually broke this bond and *irremediably* lost their status. This assumption, often retold in terms of a fall, is read by some psychoanalytical schools as a metaphor of the rupture of the symbiotic relationship between the child and the mother. At any rate, it puts the self within two strong and mobilizing tensions: that of nostalgia and that of victimhood. From the viewpoint of *nostalgia*, the Western self knows (and is constantly reminded by religious agencies) that he once belonged to another state, a more blessed one, in which its individuality was subsumed into a blissful harmony (Tannock 1995). Caillois (1935) has written of the unconscious drive of man toward a blissful immobility; and his concepts are not too far from a romanticized version of the bleak *Todestrieb* we find in Freud's later writings (i.e., Freud 1922). Of course, instead of this blissful immobility, human beings are condemned to *move* to survive, to affirm their individuality, and by moving they must perceive the limitations of physicality. Therefore, to survive, human beings have to *suffer*. This act of "necessary movement" is often symbolized by work. Indeed, the Old Testament makes it explicit that work is a *curse* God put on Adam and Eve. The self is therefore nostalgic about an Eden that is forever lost (Jacoby 1985); at the same time, it is locked in a constant struggle with a hostile nature. Therefore, man is the constant victim, always in peril, and always looking for means of survival. We may locate in this double tension (victimhood and nostalgia) the roots of the modern idealization of a-physicality, of exemption from *bodiness*, an ancient dream of which every epoch and context produced cultural representations, up until our own "disembodied identities" metaphors. However, cultural forces up until the late Renaissance only allowed this dream to be dreamt by one (or at best by a chosen few), namely, the ruler, who enjoyed the privilege of the *palace*, that is, of a megamachine (Mumford 1967) made of both organic and mechanical parts, designed to relieve them from the *necessity* of physical movement. In its ideal representations, the god-king enjoys the privilege of moving only when he *chooses* to. From this perspective, among the many functions of the palace/machine, we can add the symbolic function of granting ruling classes the utmost privilege of approximating the Edenic state. To some extent, the dream of machines as servants and of machine worlds can be grounded in the dream of the palace/machine.

When millenarian tendencies brought about the idea that a new Eden was at hand, and would come about as soon as man had prepared the conditions for it, Western man started collectively looking around for the means of building Eden on Earth. Utopian thought took it upon itself to set the blueprint for such a project. At first, political and moral engineering seemed to be the only way to go, as in Thomas

Moore's seminal work *Utopia*, which is largely unconcerned about technology in its attempt to build the ideal society. And while the utopian works of Tommaso Campanella and Roger Bacon do include the descriptions of a few wondrous machines in their cities, it is only with Francis Bacon's revolution (as epitomized in the *New Atlantis*) that science and technology become prime instruments for a cosmic change. Francis Bacon states a fundamental tenet of contemporary Western thought: that to built Eden, nature must be broken, subdued, and dominated *once for all*, and the only way to do this is by putting up an institutionalized system of science and technology (Bacon 2010). A new man can be brought about; nostalgia can find fulfillment. Man will be able to regain Eden by *rebuilding its conditions*. Man will be freed from the toil of work, and, eventually, the burden of the body and its limitations will be overcome. It is impossible to overstate the importance of this promise for the cultural framing of technology in Western thought. Following Leo Marx (2000), we can say that, once framed as purveyors of a new Eden, machines became *sublime* objects, through which to contemplate in awe *and* terror the immense, overwhelming power of technology: the power of bringing forth nothing less than a *cosmic renovation*. From this perspective, given that the feeling of the sublime evokes both awe and terror, there is no difference between those who say that machines will save us and those who say that machines will destroy us. The real difference is between those who are *enchanted* by technology and those who are *not*. The former acknowledge technology as an overwhelming power, whether experienced in awe or in terror; the latter just does not care, and might approach technology in a purely instrumental or pragmatic way. Hence, the technological sublime, a form of emotional participation that lies at the core of symbolic investment in technology, irrespective of the specific form technology may assume in a given historical moment. From this perspective, it is unsurprising that each "new" technology seems to carry along the same transformative promises of the previous ones (the end of time, of place, and of politics – see Mosco 2004); as Mosco and Carolyn Marvin showed (Marvin 1988), the current high symbolic investment in information technologies is only the contemporary form of an enchantment that had previously experienced the sublime through the railway, the telegraph, radio, and television. As many scholars noted (Mosco 2004; Nye 1996), when discourse about technology is informed and inspired by the technological sublime, it assumes the form of *myth*.

Techno-myths draw on the inspirational power of the technological sublime to mobilize for or against technology, be it in specific or even general terms.[1] As myths, such narratives are held as unconditionally true by their believers, encourage emotional participation over rational reckoning, inspire individual practices and public policies, inform processes of meaning attribution, and are re-elaborated by popular

[1] Good examples are Nicholas Negroponte's editorials in the American technology magazine *Wired*. Consider the following passage about the transformation of politics in the digital era: "Nations, as we know them today, will erode because they are neither big enough to be global nor small enough to be local. The evolutionary life of the nation-state will turn out to be far shorter than that of the pterodactyl. Local governance will abound. A united planet is certain." See Negroponte (1998). Being Global. *Wired*, 6.

culture in the forms of folktales (such as movies, novels, or comic books). More importantly, myths, as Roland Barthes wrote, tend to be incorporated into common sense, to become invisible but very much working parts of our symbolic world (Barthes 1957). In a world of quick interconnectedness, it is no wonder that such narrations enjoy wide transnational circulation. But we must not forget that these myths are the product of a very specific cultural framing of technology, which was operated by a specific culture in a specific moment in history. Hence, the question underlying our research: have other cultures produced their *own* technological sublime? And more specifically, to use a term coined by Vincent Mosco (2004), is there a Chinese *digital* sublime (i.e., specifically related to computer technology)? The thesis we will explore in the next paragraphs is the following: until the Opium Wars, Chinese culture was either unconcerned with or critical of machines. During the Century of Humiliation, the Sinocentric symbolic universe of the elite was shattered, and China's root image switched from that of the self-sufficient hero to that of the victim, therefore mobilizing a vast array of intellectual and material forces scouting the outside world for a regenerative agent able to help China *survive*. Among them, science and technology emerged as central. Thus, the machine was framed as a sublime object, able to bring forth a cosmic change; but *that change is very different* from the one promised by Western culture.

3 Roots of the Chinese Technological Sublime

Mark Elvin (1975) has observed that the "practical" attitude of classical Chinese culture toward its machines forbade the development of an "imaginative attitude" toward them "comparable to Leonardo's." The Machine hardly ever entered the realm of fancy; it was treated as a mere tool. This "practical" nature had both a cultural and material basis. Within the framework we sketched in the previous paragraph, we can observe that from both points of view, for a long time, China simply did not *feel the need* to frame the machine as a sublime object. From the cultural point of view, Chinese cosmogony features no myth of the Fall: there is no harsh dichotomy between man and nature, and the relationship between heaven, man, and Earth can be kept in harmony through moral practice. Therefore, no need emerged for an external "technology of salvation," and throughout history, laborsaving machines have been seen pragmatically by Confucians (who, privileging the inside world of man to the outside one, often dismissed the question of technology as unworthy of attention) and suspiciously by Taoists (both Laozi[2] and Zhuangzi[3]

[2] "*If a country is governed wisely, its inhabitants will be content. They enjoy the labor of their hands and don't waste time inventing labor-saving machines.*" Quoted in Mitchell (1994).

[3] See the famous fable about the irrigation machine quoted in Giles (2005: 147).

indicated machines as unnecessary "deceivers" of nature and vehicles for desires[4]). From the material point of view, according to Elvin's theory of the "high-level equilibrium trap" (Elvin 1972), the wide availability of inexpensive human labor supply made the mechanization of production simply useless; hence, raising labor productivity through mechanization was considered less important than innovation in agriculture (Chan and Goodman 1999).

This having been said, it would be equally mistaken to impute to China an inherent *distrust* of technology. Elvin himself observed that "in China there was evident enthusiasm for machines as well. Crowds turned out around 1,800 to see a new water-pump, and poetry contained references to machines," and further on reported the examples of the distich about automatic textile machines included in Wang Zhen's (1290–1333) *Treatise on Agriculture* (Nong Shu [农书]1313):

> It takes a spinner many days to spin a hundred catties/But with water power it may be done with supernatural speed.[5]

Again, Elvin remarks on Wang's "keen delight with the automatic nature of the process" in the following verses:

> There is one driving-belt for wheels both great and small;/When one wheel turns, the others all turn with it./The rovings are transmitted evenly from the bobbin-rollers./The threads wind by themselves onto the reeling-frame.[6]

More examples from the Sung, Ming, and early Qing eras could be added. Comparative studies in Chinese literature show that machines were actually *more* present throughout Chinese poetry, from the Sung Era onward, than in premodern European times, when poetry was more concerned with romance and adventure (Schmidt 1994).

Therefore, we can conclude that throughout the material, political, and cultural history of China, labor-saving machines have been known, developed, employed, and appreciated. However, what they lacked was the metaphysical dimension that made them sublime in the West – that is, the idea that they could be *more* than means to immediate ends.

It took the cultural trauma of the loss of national integrity to change this order of things. The period between the first Opium War (1839) and the end of the Japanese invasion (1945), during which China's territorial integrity was repeatedly violated, is known as the *Century of Humiliation*, and, as many commentators observed, it

[4] Taoist tradition – especially the *Zhuangzi* – holds that a "machine heart in the bosom" impedes purity. Notice how the word used here is Jixie ("machinery"), which stresses the "complexity/cunning" element more than the often-used Jiqi ("machine"). For an inquiry into the various terms used to designate the machine in Chinese and their historical significance, see Zhang 2001. The Taoist conception of technology cannot, however, be reduced to simple forms and is dynamically connected with the concept of *Wu Wei*: for a brief summary of this complex topic, see Parkes (2003).

[5] Quoted in Elvin (1975).

[6] Ibid., 199.

has been elaborated by modern Chinese culture as a defining trauma (Gries 2004). After centuries of perceived homeostatic cultural integrity (certainly dynamic in its continuous rearranging but nonetheless stable in its core nature), the loss of sovereignty at the hand of the British in 1842 and even more so at the hand of the Japanese during the Sino-Japanese war, along with the subsequent catastrophes (including the Taiping rebellion, the second Opium War, and the Sino-Japanese conflicts), brought about a change in the fundamental Chinese root image (and the founding narrative of national identity) which switched (although in a dynamic and highly complex way[7]) from *hero* to *victim*. As remarked by Gries, this was a monumental change for Chinese self-perception and is still very relevant today:

> The "century of humiliation" is neither an objective past that works insidiously in the present nor a mere "invention" of present-day nationalist entrepreneurs. Instead, the "century" is a continuously reworked narrative about the national past central to the contested and evolving meaning of being "Chinese" today (Gries 2004: 47).

As a victim, China experienced a loss and started to search for the means for salvation. The machine (embodied by Western technology) appeared as such – albeit not the only one[8] – to some of the cultural and political elites of the time. A good number of both Neo-Confucian conservative thinkers and May Fourth iconoclasts believed that technology and science would have saved China. However, it did not escape these reformers that in order to successfully introduce S&T in China, social and cultural reforms were needed. One of the imperatives became "national salvation through science."[9]

"Traditional nationalist" reformers such as Zhang Zhidong (1837–1909) and Kang Youwei (1858–1927) advocated the adoption of Western technology as a means for national salvation. Zhang Zhidong observed that "in order to render China powerful, and at the same time preserve our own institutions, it is absolutely necessary that we should utilize Western knowledge" (Snyder 2003). Zhang applied the dichotomy between *Ti* and *Yong* (roughly translatable as "essence" and "practice" or "substance" and "function") to justify the adoption of Western technology (the "practice") while holding Chinese cultural values (the "essence"). While far from impenetrable to intellectual critique (Shen and Williams 2005), this concept allowed for technology to be popularized as a means for the salvation of China while leaving more or less intact the faith in Confucianism (Kwok 1965). Kang Youwei went even further.

---

[7] Gries (2004: 51) analyzes the complexity of the relationship between "hero" and "victim" and comments on the two possible interpretations of the popular phrase "*luohou aida*" ("backward, beaten") used to summarize the century: the first, "*the backward will be beaten*" ("luohou jiuyao aida"), individuates the reason of the beating of China in its technological backwardness. The second, and presently more popular, possible interpretation, "*backward because beaten*" ("luohou yinwei aida") stresses the cause of China's late development as being Western aggression.

[8] Throughout Chinese history, social and cultural engineering has been preferred over material technology. Zhang Zhidong himself acknowledged that Chinese "proper" science is the "science of the soul."

[9] In Chinese characters, "科学救国."

Possessed by what some commentators described as an almost messianic ethos concerning his "mission to save humanity" (Chesneaux et al. 1976), Kang, like Bacon, popularized his "salvation program" through literary utopia. His work *Da Tongshu* ("The Book of Great Unity") is the description of an ideal "world society," where technology and mechanization play integral parts in granting a better life through the emancipation from work (as people in Kang's utopia only work 4 h a day). For the first time, labor-saving through machines is presented as an *ideal*.

A few years later, from another political side, anarchist Wu Zhihui (1865–1953) explicitly spoke of "saving the world through machines" as a means of granting Zhuang Zi's ideal life of "free and easy wandering." In this life, people would be freed from work and disease, by reaching a hyper-connected state "where everywhere becomes nowhere" – through technologies such as "great ships, trains, balloons, and sea diving bells, even mechanical conveyor belts for commuting between settlements" (quoted in Twitchett and Fairbank 1978: 398).

Another prominent late-Qing intellectual, Tang Caichang (1867–1900), described (using mechanist terms, a literary device also employed by Kang Youwei) the cosmic change promised by the machine in its appeal for the mechanization of Chinese industry and agriculture:

> When we make universal use of machines, then the affinity (aili) will be bright and beautiful; when affinity is bright and beautiful, then the force of the country (guoli) will become stable; when the force of the country becomes stable, the expansive force (zhangli) and the gravitational force (sheli) will be strong enough and then, even if every country attracts (xi) [China], it will not be swallowed.[10]

Hu Shi (1891–1962) made one final bold step and turned the opposition between "Western materialism" and "Chinese spirituality" upside down by provocatively stating that the Western civilization, which freed men from the toil of work through machines, was actually *more spiritual* than that in China, which "used human beings as beasts of burden" (Hu 1929). Hu thus acknowledged the value of the machine as an agent of cosmic change, able to bring forward a new human condition.

The five thinkers we quoted as examples came from different political backgrounds, and more could be added. While such technological enthusiasm would taper off, and the emancipation from work would no longer feature in fantasy, what would henceforth be engrained in the imagination of Chinese elites was the concept that science and technology could be effective agents in exiting their status as the nation of victims and restoring the original status as a hero. The notion that "China must *again become* strong" constitutes the *fil rouge* of modern Chinese politics, irrespective of ideological orientation (Gries 2004). Therefore, rapid technological development became a key part of the PRC political agenda from the early 1950s and was given a decisive boost after Deng Xiaoping's reforms (Lu 2000; Maier 1988a, b).

[10] Quoted in Lackner et al. (2001).

This is the core of our hypothesis, that the Century of Humiliation created the conditions for a reframing of the machine as a "sublime" object – that is, the acknowledgement of its potential as an agent for a cosmic change. But this change was never to make a king of the *individual* by allowing him to transcend his earth-bound limitations, like in the West; the promise of this sublime technology was to allow China *as a nation* to recover its status as a hero, to shield itself from external aggression, and eventually regain its status as a superpower.

We shall now proceed to examine the shape this peculiar form of technological sublime has taken in the Chinese collective imagination by drawing on science fiction literature, which many scholars described as a privileged place – at least in the West – for the observation of the dynamics relative to the cultural representations of technology (James and Mendlesohn 2003; Ben-Tov 1995).

4 Imagining a New Machine: Chinese Science Fiction

Chinese science fiction is often described as a minor genre, both in terms of quantity and quality of output. While this is certainly true for cinema (for science fiction is conspicuously absent),[11] regarding printed paper, the situation is more complex. When not subject to political condemnation (and subsequent crackdowns), science fiction has actually enjoyed good diffusion in absolute terms (Tidhar 2003), which does not necessarily mean, of course, that it has made any significant cultural impact. At any rate, the authors of Chinese science fiction have been from the very beginning acutely aware of the social significance of their work as manufacturers of what could be defined as "interfaces" to the new Chinese technological sublime: texts that allowed their fellow countrymen to access the inspirational power of the sublime.

The first works of science fiction in China came from late-Qing intellectuals and were mostly translations of Victorian authors such as Jules Verne.[12] However, the genre soon evolved into a fully Chinese endeavor, which David Wang labeled as "Science Fantasy" (Wang 1997). The aim of this genre was to allow a first mass celebration of the Chinese technological sublime. As noted by Tsu (2005):

> Supplementing the intellectual social utopia constructed up by Kang Youwei and Tang Caichang, futuristic fiction spells out the narrative of the yellow race's domination in no uncertain terms. The newly emerged genre of science fantasies in the late Qing provided the medium through which Western Technology and Chinese hegemony could become one. [...] Science fantasy grafted onto the materiality of Western military power the imagination for a future of Chinese supremacy.

---

[11] Among the few exceptions, we can include the movie "*Composite man*" (*He Cheng Ren* 1988).

[12] See for example the 1902 translation of *Twenty-Thousand Miles Under The Sea (*published in Ling Qichao's literary magazine *Xin Xiaoshuo*) and Lu Xun's translations of *Journey to the Moon* (1903) and *Journey to the Centre of the Earth*. In the preface of the former, Lu declared his aim being introducing modern science in China "under the guise of entertainment."

After Tsu (2005), we will illustrate this point using the example of one among the first original science fantasy tales in China, *Electrical World* (*Dien Shijie* 电世界), written in 1909 under the pseudonym of Gaoyang Bucaizi. It tells the story of a reinvigorated China of 2010, now among the world's most powerful nations by the force of its army, science, and technology. It tells the tale of the struggle of its godlike scientist protagonist, "Electric king" Huang Zhenqiu ("*the yellow that shocks the earth*"), to build "a new world" designed by China, using electricity. In the first chapter of the novel, he builds the Electric Empire Hall, a temple for the mass worshipping of the technological sublime, and where all kinds of technological novelties from electronic music to new means of transportation are experienced. The exposition is centered upon one particular relic (an energy-producing asteroid fragment) which mesmerizes those who see it "as if electricity was sucked out of their pupils." Using his electrical power (which allows him to fly), in the central part of the novel, Huang single handedly (and quite reluctantly) annihilates an entire enemy nation (the Mighty West, lead by Napoleon X) and builds a new, hyper-technological ideal world, a world of Great Unity in the vein of Kang Youwei's utopia, dominated by China. The *Dian Shijie* is full of sublime images, centered on the idea of a godlike technology able to make China again the *Middle Kingdom*, again at the center of the world. As with philosophical instances, the sublimity of the cultural representations of technology would cool down during the twentieth century, as techno-myths got incorporated into common sense, therefore reaching a much more influential position – what David Nye called "the woodwork" – over processes of meaning attribution.

Science fiction works were also published during early Maoism, as China needed to attain "socialist modernization" through technology. Therefore, science-fiction authors abandoned the "fantasy" element of the late-Qing period and adopted the Soviet model of science fiction, centered on the glorification of the scientist. As noted by Link (1987), "scientist-hero" fiction was officially encouraged by institutions as part of the modernization drive during the 1950s and 1960s. However, it enjoyed little commercial success (Wu and Mallan 2006). Furthermore, its circulation abruptly came to a halt in 1965, as a result of literary policy during the Cultural Revolution (Chong 2002). However, when, in the late 1970s, the Four Modernizations theory replaced class struggle as the inspiring principle of China, science fiction gained again a relatively prominent role (e.g., *Science Fiction World*, the world's most successful science fiction magazine, with an estimated readership of one million, was founded in 1979) (Wagner 1985), only to be banned again as a result of the "spiritual pollution" campaigns which repeatedly hit China in during the 1980s (Link 1987). Science fiction enjoyed a resurgence in the early 1990s, when, according to science fiction author Han Song (Quoted in Song 1997, emphasis ours), "The Chinese government has attached a greater importance to science and technology. The national policy of 'invigorating the country by promoting science and education' is regarded by some commentators as an engine for Chinese SF to take off. Actually, *the country needs SF* because of both political and economic reasons." Like the late Qing novelist, Song underlines the social significance of his work as an interface to the technological sublime, contributing to the "rescuing" of China: "Writing SF in

China is not a mere personal affair, but is connected with the efforts to save an economically-backward nation."[13]

In general, in Chinese science fiction one finds a much more positive outlook on science than is found in Western sci-fi. As a Chinese science fiction author observed:

> [...]Our [Chinese] science fiction is naturally different [from the Western one]. We sing the praises of science, of the glorious future for human life which a highly developed science will extend, of the various beautiful things the laboring masses will create through their reliance on science, and of the heroic struggle of our one billion people to realize the Four Modernizations. This is why while science fiction usually describes the future, its basic reasons for doing so are highly significant for the present. (quoted in Song 1997)

Because of this different nature, as noted by Doar (1982), Chinese science fiction is characterized by the absence of a major theme of Western science fiction, the diabolical scientist. This figure, which can be traced back to the myth of Prometheus, Goethe's Faust, or Mary Shelley's Frankenstein, personifies a certain form of technological terror which identifies in technology a potential for apocalypse (James and Mendlesohn 2003), is almost nonexistent in mainland Chinese literature.

These cultural representations help us to frame the specificity of the Chinese technological sublime. Its key representation is, as in the Western case, the idea of technology as an irresistible force of liberation. The difference lies in the object of this liberation. It is not the individual, longing to recover an original divine nature, transcending the limitations God condemned him to. Instead, it is a *nation*, nostalgic about its status as "Middle Kingdom," longing to recover through technology a purely immanent role in the world by escaping the status of victimhood. This has an important consequence at the level of cultural representation. In the Western technological sublime, the transcendence of the bodily limitations is always morally ambiguous, for such limitations are transcendentally rooted. Therefore, the "terror" dimension of the technological sublime often deals with this notion of "playing God" through using technology and of the catastrophes that will strike down the man who tries to overcome his earthly, limited nature. The Prometheus archetype, as incarnated in the legend of Faust, is in this respect a prime example. In China, the victimhood status is not grounded in any transcendental roots; it is a product of history, and therefore not only *can* it be reverted but it *must* be. Indeed, the sin would be *not* to. We believe this plays a fundamental role in the overwhelmingly positive attitude toward technology which can be found in the Chinese literary imagination. But if the sublime is both terror and awe – where is the terror in the Chinese technological sublime? We will discuss this in the last paragraph, where we zoom in more closely on ICTs and the Internet. Our choice of this specific technology has historical reasons: after Deng's reforms, computer technology supplanted heavy industrial machinery as the technology destined to propel China in a new era (Damm and Thomas 2006). As noted by Mueller and Tan (1996), a significant role in this

[13] Wagner (1985) called late 1970s Chinese Science Fiction "lobby literature" insofar as it was written by scientists themselves and directed toward the government asking for "a self-regulating science republic."

shift was actually played by a Western techno-mythmaker (as Vincent Mosco described him). Alvin Toeffler's neo-millenarian theories of the "third wave" presented ICTs as the necessary precondition for a society to become an "information society," the next evolutional phase after the agrarian and industrial societies. This notion became highly influential in post-Maoist Chinese elite circles. Toeffler's ideas about a power of technology that he often described as liberating for the individual (Toffler 1981) were taken, domesticated, and refitted as a part of the Chinese myth of rebirth through technology.

5 Barbarians at the Gates: Chinese Technological Sublime and Computer Technology

Having briefly sketched the philosophical roots and cultural representations of the Chinese technological sublime, we shall now proceed to our concluding movement by going back to where we started, that is, by examining its implications for the processes of meaning attribution toward ICTs. However, some preliminary warnings are in order. It has to be kept in mind that what we accounted for in the previous two paragraphs are thoughts and works of the *elite* of Chinese society; as such, they do not necessarily represent the feelings of common people.[14] Therefore, the Chinese technological sublime may be, in itself, a useful tool when applied at the level of elite practices. It might contribute to explain, for example, the apparent contradiction of an authoritarian government investing heavily in the Internet, a technology that the West routinely perceives and represents (by drawing on the experience of its own technological sublime) as intrinsically freeing and empowering (Mosco 2004). On the other hand, the Chinese technological sublime cannot be applied with equal ease to the level of the individual relationship with technology. While a certain degree of circulation between the symbolic capital of the elite and that of the masses is to be expected, it would be improper to bluntly use the Chinese technological sublime to explain how Chinese individuals give meaning to their experience of technology. It would mean to naively hold that, since a change can be recorded in "Chinese culture" after the late Qing, then the collective unconscious of all Chinese people would almost magically adjust to it. The fact that the Chinese system of thought gradually allowed for the emergence of a specific form of symbolic experience of technology simply means that the contemporary Chinese subject has now the *possibility* of accessing this experience, where previously it was possible only to relate to technology in terms of dry pragmatic utilitarianism.

To investigate at the individual level the role played by the Chinese technological sublime in the processes of meaning attribution, we conducted research in Beijing,

[14] For a critique of the dichotomy between technologist and citizens as a hindrance for China's technological development, see Shen and Williams, op. cit.

with the fieldwork taking place during the last quarter of 2007. Our team interviewed a sample of 320 subjects, asking them to describe in detail what positive and negative transformations they perceived computer technology exercised upon themselves and society. The findings indicate that at the center of the samples' symbolic universe lies a core of two interconnected elements. On the one hand, computer technology is described as a purveyor for various forms of prosperity, "saving resources" by making preexisting practices "more convenient." The positive transformations induced by technology (both at the individual and social level) appear to be of a highly pragmatic nature; indeed, most interviewees seemed not to be able to separate positive transformations from their own practices of use. Interviewees sang the praises of reading the news without the need for buying newspapers, performing instantaneous long-distance communication for a fraction of the price of phone calls, speculating on the stock market while at home (a surprisingly diffused practice in the sample), and enjoying movies, TV programs, novels, and comic books, all at no cost. Computers bought by low-income families were considered as long-term investments (sometimes almost as fetishes) for the future careers of their children. Furthermore, a large part of the sample acknowledged the important role played by computer technology in industrial and economic development.

On the other hand, there appears to be a price to pay for this prosperity. In our sample, computer technology is deemed as fostering various negative transformations upon society. It is accused of "making Chinese calligraphy disappear," by rendering this art useless. It is accused of changing the dynamics of romantic relationships by disembodying the courting phase through online dating. It is widely accused of poisoning the minds of the youth through the unchecked circulation of obscene or dangerous content. It is looked upon with suspicion as a place for public discourse, as the sample appears to be well aware of the risks of "digital lynching" connected to the use of the Internet as a "human flesh search engine." Finally, and more importantly, it is accused of being the vehicle for the disconnection of the youth from society, through the feared "Internet addiction." This last transformation, the most widely described in the sample, deserves to be discussed in detail. Our interviewees consistently indicated the Internet, and especially Internet gaming, as becoming "addictive" when one indulges in it too much. They told tales of children neglecting school and family to focus only on gaming, becoming social rejects and sometimes delinquents. They described in great detail the Internet cafes as dark, brooding, unsafe places, where, as one interviewee put it, "*children can do whatever they want, including drinking and smoking.*" More interestingly, our interviewees also described a *physical transformation* that the body would undergo when overexposed to computer technology. They described the skin as becoming "pale and wrinkled," the eyes, blind, the posture becoming crouched, while the muscles would waste away. In this representation, moral and physical degeneration appear to go hand in hand. Computer technology is represented as producing an individual that is weak, selfish, evil, and disconnected: what might be defined as a *ghost.* For a culture whose construction of the self revolves around relation and connection (Tao 1996), this ghost is the embodiment of the catastrophe.

The dynamic among the elements of this symbolic core seems to indicate the presence, in contemporary Chinese culture, of a technology-induced fear related to identity. Computer technology is perceived as being to various extents *necessary* for contemporary Chinese society. At the same time, it is perceived as a formidable *threat* to the moral education, family life, schooling, career, and physical well-being of the youth. In other words, it is perceived as damaging all of the fundamental structures the Chinese relational self is built upon. In this sense, the reference to the loss of Chinese calligraphy, one of the highest aesthetical representations of the Chinese national identity (Yen 2005), is epitomic.

Here, our findings seem to indicate, lies the core terror of the Chinese "technological sublime." Computer technology, the current bearer of the century-old promise of national rebirth, is perceived as a *barbarizing technology,* producing non-Chinese subjects, radical *others*, by undermining Chinese deep cultural structures. When examining the specific cultural representations of these fears, the parallels with late Qing elite discourses on opium, also a "foreign import" accused of destroying Chinese identity and producing ghosts,[15] appear too seductive to be ignored. Indeed, opium is a key metaphor in the elite discourses about this technology-related terror: Chinese antigaming activists such as Zhang Chunliang have been increasingly referring to Internet games as being the modern equivalent of opium. In this mythical narrative, this "digital opium" is "smuggled" into China from the West and "poisons" young generations, thus endangering the very future of China (Golub and Lingley 2008). Within our framework, this can be seen as a complementary myth to that of national rebirth, just like in the West the "digital Eden" is complemented by the "techno-apocalypse."

Our findings also appear to indicate an interesting process involving the elite and the "folk" levels of symbolic investment in technology. The sublime images elaborated by the elite (philosophers, novelists, policymakers) appear to pass through a process of "symbolic domestication" before reaching the folk level. The wonder of national rebirth is domesticated as a pragmatic "prosperity," and the terror of loss of collective identity, of the "opium," as a galaxy of microrepresentations of weakening and cultural loss. While the elite myths can be recomposed by piecing together the single or few shards possessed by individuals, none among our interviewees, no matter how high their educational level, has expressed the national rebirth or digital opium myths in a full and coherent way.

---

[15] Consider the following paragraph, in which Lo Yun Yuen describes, in 1935, opium addicts: "[All] addicts of dangerous drugs can be easily recognized by their appearance. Their complexions are grey and pale; their appearance derelict and weak. [They] grow old precociously; [their] hair turns grey easily, their pupils are contracted, teeth loosened, skin dried, [and] their voices quiver. Thus, disparaging people always call them 'opium ghosts', which is a very accurate description. [...]They spend all day lying on bed smoking opium, without doing anything for their career; their meaning of life is completely shattered by poppies." Quoted in Ho (2005: 123).

6 Conclusion

By the concurrence of a number of factors – material, historical, cultural, and social – China and the West framed the technological sublime, one of the main sources of emotional involvement in technology, in different ways. This should be taken into consideration when comparing the cultural meanings given to ICTs in the two contexts. The Western technological sublime is framed as a force that promises the emancipation of the individual from the limits connected with human nature. Therefore, the Western cultural relationship with technology focuses on the *empowering* of the individual as a means of breaking free, transcending, and escaping the perceived boundaries of the human condition. In China, the technological sublime has been framed as a force promising a national, collective rebirth. Therefore, Chinese culture appears to be less prone to experiencing the awe of self-empowerment (although using the same technology and just as extensively as the West). When imagining the transformations brought about by technology, Chinese culture appears more concerned with pragmatic small-scale improvements. And while the terror of Western man appears to be related to the uncertainty that lies beyond that same boundary breaking he is in awe of, the Chinese man appears more concerned with the loss of "Chineseness" that may be the price of a national rebirth. If the fear of the Western man appears to be loss of control, that of the Chinese man appears to be the loss of himself.

Acknowledgments This research benefited enormously from the active cooperation and invaluable discussion provide by a class of postgraduate students from the Communication University of China in Beijing, coordinated by Dr. Eileen Luo Qing. The team was comprised of Betty Qian Zhan, Cao Li, Hannah Lu, Claire Luo Dan, Gong Wiliang, Jiang Chunsheng, Li Li, Ma Jing, Wang Yi, Chang Peiyi, Xu Cong, Yang Dong, and Yi Qian.

References

Bacon F. (2010). *The New Atlantis*. Whitefish: Kessinger Publishing.

Barthes, R. (1957). *Mythologies*. Paris: Éditions du Seuil.

Ben-Tov, S. (1995). *The artificial paradise: Science fiction and American reality*. Ann Arbor: University of Michigan Press.

Bucaizi, G. (1909). Dian shijie. *Xiaoshuo shibao, 1*, 1–58.

Caillois, R. (1935). Mimétisme et psychasthénie légendaire. In *Le mythe et l'homme*. Paris: Gallimard.

Chant, C., & Goodman, D. (1999). *Pre-industrial cities & technology*. London: Routledge.

Chesneaux, J., Bastid, M., & Bergère, M. C. (1976). *China from the opium wars to the 1911 revolution*. New York: Random House Inc.

Chong, W. L. (2002). *China's great proletarian cultural revolution: Master narratives and post-Mao counternarratives*. Lanham: Rowman & Littlefield Publishers.

Damm, J., & Thomas, S. (2006). *Chinese cyberspaces: Technological changes and political effects*. London: Routledge.

Davis, E. (1999). *TechGnosis: Myth, Magic, and Mysticism in the Age of Information*. New York: Three Rivers Press.

Dery M. (1997). *Escape velocity: Cyberculture at the end of the century*. New York: Grove Press.
Doar, B. (1982). Speculation in a distorting mirror: Scientific and political fantasy in contemporary Chinese writing. *Australian Journal of Chinese Affairs, 8*, 51–64.
Elvin, M. (1972). The high-level equilibrium trap: The causes of the decline of invention in the traditional Chinese textile industries. In W. E. Wilmott (Ed.), *Economic organization in Chinese society*. Palo Alto: Stanford University Press.
Elvin, M. (1975). Skills and resources in late traditional China. In D. Perkins (Ed.), *China's modern economy in historical perspective* (pp. 95–113). Palo Alto: Stanford University Press.
Freud, S. (1922). *Beyond the pleasure principle*. London, Vienna: Intl. Psycho-Analytical.
Giles, H. A. (2005). *Chuang Tzu: Mystic, moralist, and social reformer*. Boston: Adamant Media.
Golub, A., & Lingley, K. (2008). "Just like the Qing empire": Internet addiction, MMOGS, and moral crisis in contemporary China. *Games and Culture, 3*(1), 59–75.
Gries, P. H. (2004). *China's new nationalism: Pride, politics, and diplomacy*. Berkeley: University of California Press.
Ho, V. (2005). *Understanding Canton: Rethinking popular culture in the republican period*. Oxford: Oxford University Press.
Hu, S. (1929). The civilizations of the east and the west. In C. A. Beard (Ed.), *Whither mankind: A panorama of modern civilization*. Whitefish: Kessinger Publishing.
Jacoby, M. (1985). *The longing for paradise: Psychological perspectives on an archetype*. Boston: Sigo Press.
James, E., & Mendlesohn, F. (2003). *The Cambridge companion to science fiction*. Cambridge: Cambridge University Press.
Kwok, D. (1965). *Scientism in Chinese thought, 1900–1950*. New Haven: Yale University Press.
Lackner, M. et al. (2001). *New terms for new ideas: Western knowledge and lexical change in Late Imperial China*. Leiden, Boston, Tokio: Brill.
Link, P. (1987). The limits of cultural reform in Deng Xiaoping's China. *Modern China, 13*(2), 115–176.
Lu, Q. (2000). *China's leap into the information age: Innovation and organization in the computer industry*. New York: Oxford University Press.
Maier, J. H. (1988a). Computer science and information technology in the People's Republic of China: The emergence of connectivity. In M. Yovits (Ed.), *Advances in computers*. San Diego: Academic Press Professional.
Maier, J. H. (1988b). Computing in China: The People's Republic: Building a computer infrastructure. *Abacus, 5*(2), 34–42.
Marvin, C. (1988). *When old technologies were new: Thinking about electric communication in the late nineteenth century*. London: Oxford University Press.
Marx, L. (2000). *The machine in the garden: Technology and the pastoral ideal in America*. Oxford University Press.
Mitchell, S. (1994). *Tao te ching: A new English version*. New York: HarperCollins Publishers.
Mosco, V. (2004). *The digital sublime: Myth, power and cyberspace*. Cambridge: MIT Press.
Mueller, M., & Tan, Z. (1996). *China in the information age: Telecommunications and the dilemmas of reform*. Westport: Greenwood.
Mumford, L. (1967). *The Myth of the machine: Technics and human development*. San Diego: Harcourt Brace Jovanovich.
Negroponte, N. (1998). Beyond digital. *Wired, 6*(12), 288.
Noble, D.F. (1997). *The Religion of Technology: The Divinity of Man and the Spirit of Invention*. New York: Alfred A. Knopf.
Nye, D. E. (1996). *American technological sublime*. Cambridge: MIT Press.
Parkes, G. (2003). Lao-Zhuang and Heidegger on nature and technology. *Journal of Chinese Philosophy, 30*(1), 19–38.
Schmidt, J. D. (1994). *Within the human realm: The poetry of Huang Zunxian, 1848–1905*. Cambridge: Cambridge University Press.
Shen, X., & Williams, R. (2005). A critique of China's utilitarian view of science and technology. *Science Technology and Society, 10*, 197–223.

Snyder, L. (2003). *The new nationalism*. Piscataway: Transaction Publishers.
Song, H. (1997). The social environment of Chinese science fiction. In A. Hull (Ed.), *1997 Beijing international conference on science fiction: Essays*. Chengdu: SFW.
Stefik, M. (1997). *Internet Dreams: Archetypes, Myths, and Metaphors*. Cambridge MA: MIT Press.
Tannock, S. (1995). Nostalgia critique 1. *Cultural Studies, 9*, 453–464.
Tao, J. (1996). The moral foundation of welfare in Chinese society: Between virtues and rights. In G. Becker (Ed.), *Ethics in business and society: Chinese and western perspectives* (pp. 9–24). New York: Springer.
Tidhar, L. (2003). Science fiction, globalisation, and the People's Republic of China. *Foundation: The International Journal of Science Fiction, 89*, 93–99.
Toffler, A. (1981). *The third wave*. New York: Bantam Books.
Tsu, J. (2005). *Failure, nationalism, and literature: The making of modern Chinese identity, 1895–1937*. Palo Alto: Stanford University Press.
Twitchett, D. C., & Fairbank, J. K. (1978). *The Cambridge history of China*. Cambridge: Cambridge University Press.
Wagner, R. G. (1985). Lobby literature: The archeology and present functions of science fiction in China. In J. C. Kinkley (Ed.), *After Mao: Chinese literature and society* (pp. 17–62). Cambridge/London: Harvard University Press.
Wang, D. (1997). *Fin-de-siècle splendor: Repressed modernities of late Qing fiction, 1849–1911*. Palo Alto: Stanford University Press.
Wu, Y. J., & Mallan, K. M. (2006). Children's science fiction in China. *Bookbird, 44*(3), 40–47.
Yen, Y. (2005). *Calligraphy and power in contemporary Chinese society*. London: Routledge.
Zhang, B. (2001). An inquiry into the history of Chinese terms jiqi (machine) and jixie (machinery). In M. Lackner, I. Amelung, & J. Kurtz (Eds.), *New terms for new ideas: Western knowledge and lexical change in late imperial China* (pp. 177–195). Leiden: Brill Academic Publishers.

Chapter 16
ICT Use with Chinese Characteristics

Wai-chi Rodney Chu and Yinni Peng

1 Introduction

Today, the importance of Chinese use of ICTs—especially mobile phone and Internet use—is incontestable. By year-end of 2010, the number of mobile phone users in the People's Republic of China (hereinafter China) had reached 859 million. According to the annual report of Ministry of Industry and Information Technology (MIIT) on 2010 nationwide statistics on information industry, China now has the largest number of mobile phone users in one country, but this still only represents a penetration rate of 64.4/100, implying that there is ample room for market expansion. As for Internet users, by year-end of 2010, the number had reached 457 million, of which 450 million were using broadband; yet, the penetration rate was talking about 34.3% only (MIIT 2011).

Whereas a commercial orientation would emphasize available room for further market expansion, from a sociological standpoint, interest lies in discovering what exactly did happen to this large group of Chinese ICT users when they experienced this exposure in such a short time span *alongside* the sort of compressed modernity (a term that will be further elaborated below) that the whole country has been undergoing. In relation to this, how much can the ICT experience, which has been well documented in Western literature, help us to understand the Chinese users? For instance, a standard assumption of ICT studies is that information technologies have led to even greater individualization and have enabled human beings to achieve even greater privacy (Kopomaa 2000; Licoppe and Heurtin 2002; Puro 2002; De Gournay 2002;

W.-c.R. Chu (✉)
Department of Applied Social Sciences, The Hong Kong Polytechnic University, Hong Kong, China
e-mail: ssrodney@inet.polyu.edu.hk

Y. Peng
Department of Sociology, Hong Kong Shue Yan University, Hong Kong, China
e-mail: pengyinni365@gmail.com

P.-l. Law (ed.), *New Connectivities in China: Virtual, Actual and Local Interactions*,
DOI 10.1007/978-94-007-3910-9_16, © Springer Science+Business Media B.V. 2012

Fortunati 2002). Now, what does "individuality" mean in China, or, for that matter, to the Chinese? How, for example, is the public/private divide perceived? Of no less importance, how do we account for the enormous scale and the rapid pace of penetration of ICTs in the course of a decade? By posing these questions, we are not implying that Chinese ICT users are entirely unlike users in the rest of the world. Rather, we are investigating whether there are unique effects due to Chinese cultural traits *and* the speed of diffusion of ICTs in China, effects that it will be costly to downplay. We wish to determine how the social values and behavioral norms of the Chinese influence the cyber aspects of life. We also want to gauge the impact of the rapid spread of ICTs, given this sociocultural context. Based on some general captures of fieldwork data undertaken since 2003 in Guangzhou and Beijing, this chapter tries, in a modest manner, to provide an examination on the relationship between society and technology in the Chinese cultural context and see how self, privacy, and *mianzi* (face) are perceived through ICT use in cyberspace. In addition, it also suggests how Chinese cultural traits and the speed of the ICT evolution in China may combine to bring about a unique cyber experience. This analysis may be helpful to other scholars who wish to compare the impact of ICTs in various cultures or who are interested in discovering how Mainland China is going cyber.

2 Theoretical Snapshot and Contextual Outline

The rapid development of ICTs has greatly facilitated the presence of absent others. In his analysis of late modernity, Anthony Giddens emphasizes that this period is characterized by the separation of time and space and the "dis-embedment" of social systems (Giddens 1990). As a result of the separation of time and space, interactions in modern society are no longer constrained by the necessity of presence; it has become easy to have relationships with "absent others." "Place" is increasingly phantasmagoric: an activity that is geographically situated in a particular locale can be penetrated and shaped by social influences at a distance (Giddens 1990: 19). Closely related is the notion of "dis-embedment," which refers to the "lifting out" of social relations from local contexts of interaction and the restructuring of these relations across indefinite spans of time and space (Giddens 1990: 21). Whereas Giddens focuses on the direction, rather than the content, of changes in contemporary life, Kenneth Gergen looks at its preoccupations. In his discussion of ICTs, he tags the term "absent presence" (Gergen 2002), showing how one can be physically in one location but absorbed by a technologically mediated world elsewhere (Gergen 2002: 227). Barry Wellman identifies another important characteristic of the ICT age: "the emergence of network individualism" (Wellman 2001). He argues that "the person has become the portal," maintaining that it is now the individual, and not the household or the group, that is the primary unit of connectivity (Wellman 2001: 238). Communication and community have gradually moved from "door-to-door" to "place-to-place" to "person-to-person," and social context is gradually uplifted in this process. As new technology that prioritizes and enhances interactions becomes increasingly available, the power of person-to-person communication systems is poised to increase (Wellman 2001: 241). Understandably, in the contemporary era, our main relationships will be with the "absent others" on the other end of the technological device, and they will take much of our attention and determine our moods.

With the emergence and normalization of this kind of behavior, attention has focused on the ways ICTs in general, and mobile phones in particular, extend the private sphere into the public sphere (Puro 2002; Fortunati 2002; Caporael and Xie 2003). Some studies conclude that public mobile phone conversations challenge traditional behavioral standards and require the adjustment of public perception (Kopomaa 2000; Licoppe and Heurtin 2002). More bluntly put, a consequence of the information society is that the day-to-day use of ICT facilities has challenged the public/private divide that Westerners believe they used to enjoy, and so it becomes an issue that raises particular concern.

However, we must ask if this development is considered a *threat* in non-Western settings. Put it the other way, is the concern in the West receiving the same weight of attention otherwise? For instance, Pertierra argues that the private and public dichotomy cannot be transplanted directly from the Western to the Philippine context (Pertierra 2002: 44). In the case in China, is the blurring of private and public spheres so much an issue that it deserves investigation in the Chinese context? Perhaps we should proceed with caution when using Western research to determine the sociological meaning of ICT development in Mainland China. Thus, when an increasing number of studies have informed us that ICTs can be both empowering and controlling (Kopomaa 2000; Wellman 2001; Harwit and Clark 2001; Yang 2003a, b), the concern here is to what extent these studies have taken into account the cultural factors that affect the use of these devices.

Yet, it is here that ICT study in China has to be given special attention. In the last few decades, Mainland China has undergone a sort of compressed modernity. As such, it means the country has experienced a full-scale sociocultural and techno-economic upheaval to such an extent that people of all walks of life have some sense of it. China study literature captures the increasing dilemmas—in terms of life chance, life choice, and the issues of right and good—faced by the people living through this compressed modernity (see, for instance, Chu and Ju 1993; Goldman 2005; Hertz 1998; Link et al. 2002; Lull 1991; Madsen and Pickowicz 2002 Pun 2005; Walder 1986; Yuen et al. 2004; Zha 1995). But, while China is still in the early stages of riding the juggernaut of modern capitalism, the world has stepped into the cyber era. History shows the rapid evolution of ICTs has become a global phenomenon. Even in the United States, the speed of the development of Internet and Website culture has been astonishing, while terms like cyber sociology/sociologists, cybernetic age, and network society have emerged in only the last 10 years. Whereas social scientists in the West have been busy depicting the origins, meanings, and implications of the cyber age in an already industrialized/modernized setting, in the last decade or so, China, determined to join the global world, has experienced industrializing and going cyber in an almost simultaneous manner. This is the unique situation that China faces, and it has added weight to the density of its compressed modernity. Simply put, ICT exposure has further accelerated the social dynamics in China in the last decade, and more twists and turns in the normative and behavioral aspects of the contemporary Chinese are to be expected. This phenomenon, or what Chu contends as "riding the double juggernaut" (Chu 2012), is a context that China watchers have yet to turn to. For instance, it is crystal clear that those Chinese who can afford to encounter this exposure have caught up with the latest technological advances. As such, how does the cultural and personal background of the users affect their ability to adapt to ICTs and shape the meaning of ICT use? From this, process derives concern at the rich–poor digital divide. How might the consequences of this economic division under a context of compressed development affect social relations? These questions are more than enough to persuade us of the necessity of going into the Chinese context to capture the characteristics of Chinese ICT use.

Evidence outlining the essence of the Chinese cultural context has indeed informed us on the uniqueness of Chinese culture. Whitman points out that, unlike the West, China does not put the claims of the individual above all others (Whitman 1985: 86). Privacy, which is highly prized in the West, "… [allows] an individual to exercise free choice as part of a process of self-determination that has little to do with the larger world" (Whitman 1985: 97). This, however, is not the way the Chinese perceive the role of an individual in an ideal society. Potter and Potter use the expression "Image of Irrelevant Affect" to describe how Chinese culture downplays emotion as a function of expressing one's inner feelings because the Chinese do not believe emotional expressiveness is useful in achieving one's goals (Potter and Potter 1990: 183–185). The implication of this difference between China and the West is clear: what appears to be normal and fundamental to one group may not appear so to another. The issue at stake may be captured by the term "situational self," which is a synthesis of Fei's (1992) concept *chaxugeju* (differentiated mode of association), the idea of the "relational self" as termed by King (1985), and the "situation-centered" characteristic of Chinese as coined by Hsu (1981[1953]). The situational self concerns how the Chinese perceive relationships with those around them, and from there decide the appropriate way of acting.

Our attempt here is to make use of this cultural perspective to come to terms with the various aspects of the ICT behavior of the Chinese, with particular reference to the use of the mobile phone in the current context. Fei's (1992) *chaxugeju* captures the essence of the Chinese interactions. To achieve the highest goal of Confucianism, *he* (harmony), one should act in a proper way, and this brings out the importance of *li* (courtesy). *Lun* (cardinal relationship) is central to the Confucian order, which stresses the importance of *bei* (differentiation or distinction). A person's maturity is witnessed from one's assessment of the distinction of different individuals and interacts with them according to the corresponding distance from one's self position. This is the ability of knowing how to *tui* (push or extend out), just as ripples flow out from the center of the splash when a stone is thrown into the water. This metaphor of Fei's captures the essence of the Chinese way of interaction in the form of *chaxu*: the self, embedded in social relationships and emotionally tied by the personal obligations defined by those relationships, behaves with the *li* (courtesy) the relationships require (Fei 1992). Behind this lies a belief in treating a person *not as* a person per se (i.e., as an individual on equal terms), but *according to* the relationship of that person with oneself. This is why the renowned Confucian scholar Liang Shu-ming (1987[1949]) concludes the chief characteristics of the Chinese society as *lunli benwei* (ethics-based), which means the focus of the social system of China is placed on the particular nature of the relationship between those interacting individuals. King elaborates this with the term "relational being": people are related to each other through concrete obligations that Chinese identify as *qing* (affection), *yi* (righteousness), *bie* (distinction), *xu* (order), *cheng* (sincerity), and the like (King 1985: 63–65). This emphasis on relationships explains why *mianzi* (face) is so important to Chinese culture (Hu 1944). Chinese pay special attention to the kind of respect that is given to or given by others. Fei's outline of the essence of Chinese values and behaviors has been closely sustained by the views of Hsu. Developed from his famous father-son dyad (Hsu 1968: 583), Hsu uses examples from art, literature, patterns of conduct between the sexes, the arrangement and patterns of behavior in the home, and schooling to show the different passage of Americans (who are individual-centered) and Chinese (who are situation-centered) (Hsu 1981[1953]: 20–29, 49–60, 78–94). The particularistic orientation of the Chinese toward those who have relationships with them means that people are not assessed in a universal manner, and naturally, no two people have to be treated similarly. The whole essence here is not a matter of whether the individual-centered ways of the West are of a higher order or not over the situation-centered nature of the Chinese; it is just that the two peoples are different. The situational self as such can be understood as a capture of how the Chinese interact with people surrounding them, and this is quite different from the Western perception of self, which, as Charles Taylor succinctly states, has its own philosophical, historical, and cultural roots (Taylor 1989).

Under the current analysis, the above articulation is not trying to say Chinese ICT users in general, and mobile phone users in particular, are different from other users worldwide. Rather, we are investigating how the social values and behavioral norms of the Chinese may lead to some particular happenings in their encounters with cyber devices. We regard this endeavor as meaningful for two reasons. First, in more general terms, the sociological study of ICT usage on Chinese soil is still in its infancy (Chu et al. 2012, Herold and Marolt 2011; Qiu 2009). A reflective concern at this juncture, if correctly addressed, may benefit future academic pursuits in this area of study. Second, in more particular terms, the study of mobile telephony on Chinese soil is itself a substantive research area in a specific cultural context. This discussion relies much on our fieldwork data that we have collected through interviews with new migrant workers, university students, and white collar workers in contemporary China since 2003 (Chu and Yang 2006). From the data, we depict how the situational self, which is a cultural feature of the Chinese, is still observable in the daily use of the mobile phone by contemporary Chinese alike. Thus, we bring forth the universal-particular concern of this social phenomenon in this kind of comparative research. On the one hand, we try to figure out how this group of people is situating itself in an environment where people are interacting so closely with and relying so much on technology; this is the universal aspect of the phenomenon. On the other hand, we should not forget that Chinese users, as one type of mobile phone user, are encountering the technology under their particular cultural heritage (how Chinese perceive the relationships between one another and the emitting effect of such perception on the behavioral norms in their daily functioning); this is the particular side of the phenomenon.

Our conception of ICT study is that it should not only depict the relation between a social self and a communication technology but also capture the relationship *between* social actors *through* the communication tools. As social actors, ICT users (be it mobile phone or the Internet) are embedded in their social environment and cultural tradition. For sure, the use of ICTs can affect the manifested level of their lives. However, at the deep-rooted level, the social actors have a role to play that constructs and changes the meaning. Based on this perspective, we want to examine how Chinese users interpret ICT usage in the Chinese context and, moving on from its use, how they understand themselves and their relationship with other social actors (see also Peng and Chu 2012).

3 Fieldwork Encounters

Since 2003, our research team has been collecting data through interviews with migrant workers (Chu and Yang 2006), university students, and white-collar workers in China in order to identify cultural traits in the cyber activities of the Chinese.[1] In one of our early papers (Chu and Peng 2005), we examined how Chinese mobile phone users perceive the difference between the public and private spheres and the symbolic meanings associated with ICT use which are related to considerations of *mianzi* (face).

3.1 Cultural Characteristics in Cyberspace

The following dialogue between an interviewer and a white-collar worker who lived in Europe for 2 years may help us to appreciate the uniqueness of cyber contact in China:

Q: Sometimes you may find that people talk so loud in public places that you can hear the content of their discussion with others. They do not consider how you feel about them. What do you think about this?

[1] The interviews with these three groups of contemporary Chinese took place in different places and at different times. Most interviews with young migrant workers were conducted in Dongguan City, Guangdong Province, in July 2003 and May 2004. Some of the young migrant workers were interviewed in Beijing in March 2005 and December 2006. Interviews with white-collar workers were conducted in Beijing in March 2005 and from October to December 2006. Interviews with college students took place in March 2004 and from September to November 2006.

A: I have had such an experience personally. My American friend was not happy with my behavior. In Europe, when you are travelling on public transport or you are attending public occasions, nobody will speak out loud….I remember one time my friend and I took the train and went to London, and my boyfriend rang me up. Because I was talking aloud, my friend looked at me unhappily and said, "Must you talk at this time?". He said that the people around me were all looking at me. Later, I noticed that the British did use the mobile phone on the train, but they normally talked in a soft tone. I started to pay attention my own practice, but I think the Chinese have grown accustomed to loud voices. My father answers all calls so loudly that even I feel uncomfortable. For instance, he would say, "Oh! President Zhou! I am in Xiamen now…." You don't have to let everybody know where you are, do you? But I think this is a habit because the Chinese think the louder the voice, the greater the respect.

The difference between the attitude of the interviewee and that of her American friend toward the use of mobile phones in public is the product of different understandings of the nature of the public and private sphere. The Chinese do not separate the public and private spheres according to geographical location (e.g., by talking softly on public transportation); they are much more focused on behaving appropriately to the recipient of their call (even if this involves talking loudly on the phone in public).

Impoliteness does not lie in disturbing people in close proximity; it lies in not respecting the person on the other end of the phone. The *xu* (order) of the absent-present-other (in this case, the interviewee's boyfriend) is far above that of anybody on the train. This *bie* (distinction) makes the person on the other end of the phone the top priority: the boyfriend is the main object of *qing* (affection). The Chinese bring their private life into the public eye, unlike Westerners, who are made uneasy by such displays (Wellman 2001: 238). The Chinese make full use of technology's ability to bring significant others into one's existing space. This is logical, given the Chinese cultural context: it is always important to act as a relational being and to treat others according to the dictates of your relationship.

Evidence of the way cultural traits influence ICT use is also provided by examining the effects of *mianzi* on users from various socioeconomic backgrounds. Some of the migrant workers in our survey, particularly the men, mentioned that they felt under social pressure to acquire a mobile phone.

Q: Why do so many migrant workers purchase a mobile phone?

A1: In the past, your status was proved by your suit; now, it is your mobile phone that counts.… If you are asked for your number and you do not have one, you will lose face. And you have more face if you have the latest model.

A2: If you don't have a mobile phone … you will have a sense of inferiority

As we have argued elsewhere (Chu and Yang 2006), the massive acquisition of mobile phones by young migrant workers is the result of a particular combination of cultural and technological developments. This group is much more aware than the older generation of the restrictions they face in a rapidly developing Chinese economy.

They do not see how they will be able to shake off their peasant roots, and, as a result, they are plagued by a frustration that was unknown to their predecessors. For this group of young migrant workers, mobile phones are not merely for the purpose of communication. A mobile phone represents the promise of modern technology and acts as a counterbalance to their frustrating prospects. ICTs offer them the closest approximation of life as an active agent in the cyberspace. Yet, in the course of pursuing this cyber life, they are not insulated from their cultural heritage. Not only does the mobile phone provide them with a means of maintaining relationships, developing networks, and strengthening bonds, it also functions as the basis for social competition. The attitude among college students is different. Most claim that *mianzi* is not entirely derived from the ownership of a mobile phone. Some even say that mobile phones are not as special as they once were (Peng and Chu 2012). We should not assume that *mianzi* is not important to college students. They have other alternatives to declare their status if they wish. Most of the college students and white-collar workers we surveyed have their own computers and access to the Internet; this technology is at least as important as a mobile phone in their day-to-day social lives. Here we see that attitudes toward ICTs are shaped not only by the devices themselves but also by the background of the owner.

For migrant workers, the short message service (SMS) plays a very significant role in their social life due to its symbolic implications. Through the use of SMS, they can manipulate technology in order to construct networks with others. This addresses a very important aspect of Chinese social behavior: without networks, it is difficult to relate to others. The mobile phone can help migrant workers cope with the pain of their existential predicament. More positively, ownership of a mobile phone bolsters self-esteem. After all, there is no other consumer good that, while held in their hands, can provide opportunities to demonstrate their feelings, confirm their sense of freedom, and proclaim their existence in an enormous time-space continuum. In the end, resource networks are produced and accumulated via technology, and a newly established *chaxugeju* appears to be reproducing itself through the use of ICTs (Chu and Yang 2006). ICTs are a means of empowerment for migrant workers whose lives, in other respects, are relatively deprived (Law and Peng 2006). Seen in this way, migrant workers' use of ICTs reveals that they are shaped by and reproduce cultural routes/practices in the daily usage of ICTs.

3.2 Implications of Exposure at Different Stages of Life

As the above discussion reveals, ICT users from different backgrounds (students, migrant workers, and white-collar workers) are all influenced by cultural traits in the conduct of their cyber lives. Another factor that we must address is the speed of ICT dissemination in China. The penetration rates of the mobile phone, the computer, and the Internet have skyrocketed in recent years. Given that these technologies are constantly being updated, familiarity with ICTs is now established not merely by owning a device but by reaching a certain level of sophistication in its use. It is easy to see the differences in opportunities for those who have access to ICTs and those who do not (i.e., cases of a simple digital divide). But even more significant are differences between various types of access; those who have easy and unlimited access to ICTs benefit much more than those who have only limited access. These differences are a by-product of the rapid dissemination of ICTs and reveal the importance of the social context in cyber China. Our interviews showed that university students and young white-collar workers who were given the opportunity to use ICTs from the time these devices were introduced have now got a competitive edge. We interviewed two 19-year-old first-year university students—one began surfing the Internet and talking with friends on QQ when he was still in grade 8, but the other had very limited exposure via a second-hand computer donated by her relatives. They presented a clear example of the new, more sophisticated, digital divide that has occurred as a result of the accelerated pace of ICT development. There are two issues of concern here. First, assuming the younger ones are all undergoing or have undergone tertiary education, the socioeconomic difference among them may already be more than enough to hamper the kind of social exposure and horizon broadening gained via the Internet. Indeed, the early work of our research team has stressed that the socioeconomic status of the students has an impact on the pattern of mobile phone consumption, and hence on their own self-esteem in daily living (Yang and Song 2004). Students without ICT capabilities are at a disadvantage not only in terms of the resources they can access but also in terms of their self-respect. One loses face when not living up to the expectations of others. Second, the age at which a person is first exposed to ICTs is critical. Since the mobile phone and Internet only became popular in China at the turn of the millennium, all Chinese have had a relatively short exposure. For instance, in 2008, university graduates or working youth might have had 8 years' experience working with computers and owning a mobile phone, and even less experience surfing the Internet. For this group, their introduction to ICTs began in their late teens. But students just entering university in 2008 were introduced to ICTs in their early teens. It is not difficult to imagine the difference in capabilities between children who began surfing the Internet at 10 and youths whose first exposure was not until 18. A gap of 8 years in cyberspace will produce very different skills and opportunities. This cohort difference does not only apply to students: the situation is similar among the new migrant workers.

When we began our fieldwork in 2003, we witnessed their mixed feelings when they perceived the difference between the realms of cyberspace and their relatively deprived backgrounds. In new cities, where they were marginalized and had little chance of permanent residence, ICTs provide an opportunity to manipulate technology and participate in cyberspace. This was obviously a face-saving strategy; in their real life, due to their educational and economic background, they were unable to control much. The research we conducted only a few years later revealed an enormous difference in this group. There are now great discrepancies among migrant workers themselves. Some are still hoping that their lives will change as a result of SMS. A small number, however, have learned to participate actively and brilliantly in the forums of the cyber universe (via BBS and blogs) at Internet cafes.[2] Reading their blogs, we can see that the emergence of these new cyber platforms provides migrant workers with an extraordinary opportunity to broaden their horizons. The impact of this phenomenon on their self-perception and the events that surround them cannot be underestimated. We must admit that the new migrant workers (of 2007) are much further ahead than those we encountered in 2003.

4 Concluding Remarks

To recapitulate, even though this analysis was limited to the use of two ICTs—the mobile phone and the Internet—we hope to form some preliminary hypotheses regarding the dynamics of cyber China. Although we have not reached the stage where we can develop a model for the Chinese ICT use, we do want to help figure out the unique side of the Chinese cyber experience. The issue at stake concerns the attitudes toward public and private spheres in China. Whereas Westerners often feel uncomfortable when the two spheres are blurred by technology advancement, the Chinese seem to lose no time in celebrating the emergence of "absence presence." In depicting this scene, we try to show how the existence of the situational self may explain certain behavioral patterns of the Chinese in the cyber world. As such, it should be a worthwhile agenda in the future to keep track on how Chinese culture may transform itself in a cyber era to which China is fully committed.

Coexisting with this cultural concern is the issue of the rapid dissemination of ICT usage in China when the country itself is undergoing a modernization process at hyperspeed. Here, we are talking about the density of change: since the opening up of China in the late 1970s, sociocultural and techno-economic upheavals have swept the country in a very short time span, and such a compressed modernizing process is hardly matched by the experience of any Western counterpart of the contemporary era. This makes the emergence of ICT usage in China unique, and also explains why the digital divide may have a more refined content. For sure, for those persons who can afford cyber exposure, the chances to embrace each new wave of technological development are never lacking. And for the more marginalized users, the feeling can be bitter when they know, but cannot reach, what is there. Yet, some social marginal groups (in this case particularly the migrant workers) may make use of the ICT environment to turn the tide, seeking psychological pleasure and/or building up cyber relations which, in such a fast-changing society, may bring to

[2] In March 2007 in the field site of Dongguan City, the research team came across two young migrant workers who had created their own blogs. The first one was a 23-year-old woman who had developed her blog at her privately rented small residence; the second was a 19-year-old female who had developed her blog at an Internet cafe. It is clear that as migrant workers in Shenzhen quite a sum is needed to support such a hobby, but it is telling at the same time that they use it as a means to express their feelings and set up their cyber link.

them a glimpse of hope for life betterment. Thus, the pace and depth of diffusion of ICTs in a China which is modernizing itself may bring about deprivation, chances, and real and false hopes that may further affect the social dynamics in a positive or negative direction in this unstable era. This, again, deserves further effort of investigation. All in all, while we have not been able to sum up all the characteristics of Chinese ICT use, the phrase does serve as a reminder that we should not underestimate the role of cultural traits and accelerated dissemination in the development of the Chinese cyberspace. Given that ICTs are evolving more quickly than ever, we must recognize that our understanding of self and society in Chinese culture should be constantly revised as a result of ever-changing relationships between culture and technology.

References

Caporael, L. R., & Xie, B. (2003). Breaking time and place: Mobile technologies and reconstituted identities. In J. E. Katz (Ed.), *Machines that become us: The social context of personal communication technology* (pp. 219–231). New Brunswick: Transaction.

Chu, W. C. R. (2012). Riding the double juggernaut: Depicting the Chinese mobile communication context. In W. C. R. Chu, L. Fortunati, P. L. Law, & S. Yang (Eds.), *Mobile communication and the greater China* (pp. 33–46). London/New York: Routledge.

Chu, G., & Ju, Y. (1993). *The great wall in ruins: Communication and cultural change in China.* Albany: State University of New York Press.

Chu, W. C. R., & Peng, Y. (2005, October). *The use of mobile phones with Chinese characteristics.* Paper presented at Mobile Communication and Asian Modernities II, Beijing, PRC.

Chu, W. C. R., & Yang, S. (2006). Mobile phones and new migrant workers in a south China village: An initial analysis of the interplay between the "social" and the "technological". In P. L. Law, L. Fortunati, & S. Yang (Eds.), *New technologies in global societies* (pp. 221–244). Singapore: World Scientific.

Chu, W. C. R., Fortunati, L., Law, P. L., & Yang, S. (Eds.) (2012). *Mobile communication and the greater China.* London/New York: Routledge.

De Gournay, C. (2002). Pretense of intimacy in France. In J. E. Katz & M. A. Aakhus (Eds.), *Perpetual contact: Mobile communication, private talk, public performance* (pp. 193–205). Cambridge: Cambridge University Press.

Fei, X. (1992). *From the soil: The foundations of Chinese society.* Berkeley: University of California Press.

Fortunati, L. (2002). Italy: Stereotypes, true and false. In J. E. Katz & M. A. Aakhus (Eds.), *Perpetual contact: Mobile communication, private talk, public performance* (pp. 42–62). Cambridge: Cambridge University Press.

Gergen, K. J. (2002). The challenge of absent presence. In J. E. Katz & M. A. Aakhus (Eds.), *Perpetual contact: Mobile communication, private talk, public performance* (pp. 227–241). Cambridge: Cambridge University Press.

Giddens, A. (1990). *The consequences of modernity.* Cambridge: Polity Press.

Goldman, M. (2005). *From comrade to citizen: The struggle for political rights in China.* Cambridge, MA: Harvard University Press.

Harwit, E., & Clark, D. (2001). Shaping the internet in China: Evolution of political control over network infrastructure and content. *Asian Survey, 41*(3), 377–408.

Herold, D. K., & Marolt, P. (2011). *Online society in China: Creating, celebrating, and instrumentalising the online carnival.* London/New York: Routledge.

Hertz, E. (1998). *The trading crowd: An ethnography of the Shanghai stock market*. Cambridge: Cambridge University Press.

Hsu, F. L. K. (1968). Chinese kinship and Chinese behavior. In P. Ho & T. Tsou (Eds.), *China in crisis: Vol. 2. China's heritage and the communist political system* (pp. 579–608). Chicago: University of Chicago Press.

Hsu, F. L. K. (1981[1953]). *Americans & Chinese: Passage to differences* (3rd ed.). Honolulu: University of Hawaii Press.

Hu, H. (1944). The Chinese concept of 'face'. *American Anthropologist, 46*(1), 45–64.

King, A. Y. C. (1985). The individual and group in Confucianism: A relational perspective. In D. Munro (Ed.), *Individualism and holism: Studies in Confucian and Taoist values* (pp. 57–70). Ann Arbor: Center for Chinese Studies, University of Michigan.

Kopomaa, T. (2000). *The city in your pocket: Birth of the mobile information society*. Helsinki: Helsinki University Press.

Law, P., & Peng, Y. (2006). The use of mobile phones among migrant workers in southern China. In P. L. Law, L. Fortunati, & S. Yang (Eds.), *New technologies in global societies* (pp. 245–258). Singapore: World Scientific.

Liang, S. M. (1987[1949]). *Zhongguo wenhua yaoyi* [Essences of Chinese culture]. Hong Kong: Joint Publications.

Licoppe, C., & Heurtin, J. (2002). France: Preserving the image. In J. E. Katz & M. A. Aakhus (Eds.), *Perpetual contact: Mobile communication, private talk, public performance* (pp. 94–109). Cambridge: Cambridge University Press.

Link, P., Madsen, R. P., & Pickowicz, P. G. (Eds.). (2002). *Popular China: Unofficial culture in a globalizing society*. Lanham: Rowman & Littlefield.

Lull, J. (1991). *China turned on: Television, reform, and resistance*. London/New York: Routledge.

MIIT. (2011, January 26). 2010 Nation-wide Information Industry Statistical Report. From http://www.miit.gov.cn/n11293472/n11293832/n11294132/n12858447/13578942.html. Accessed 4 Apr 2011.

Peng, Y., & Chu, W. C. R. (2012). Mobile phone usage in Chinese Society. In W. C. R. Chu, L. Fortunati, P. L. Law, & S. Yang (Eds.), *Mobile communication and the greater China* (pp. 189–201). London/New York: Routledge.

Pertierra, R. (2002). *Txt-ing selves: Cell phones and Philippine modernity*. Manila: De La Salle University Press.

Potter, S. H., & Potter, J. M. (1990). *China's peasants: The anthropology of a revolution*. Cambridge: Cambridge University Press.

Pun, N. (2005). *Made in China: Women factory workers in a global workplace*. Durham: Duke University Press.

Puro, J.-P. (2002). Finland: A mobile culture. In J. E. Katz & M. A. Aakhus (Eds.), *Perpetual contact: Mobile communication, private talk, public performance* (pp. 19–29). Cambridge: Cambridge University Press.

Qiu, L. J. (2009). *Working-class network society: Communication technology and the information have-less in urban China*. Cambridge, Massachusetts: MIT Press.

Taylor, C. (1989). *Sources of the self: The making of the modern identity*. Cambridge, MA: Harvard University Press.

Walder, A. G. (1986). *Communist neo-traditionalism: Work and authority in Chinese industry*. Berkeley: University of California Press.

Wellman, B. (2001). Physical place and cyber place: The rise of personalized networking. *International Journal of Urban and Regional Research, 25*(2), 227–252.

Whitman, C. (1985). Privacy in Confucian and Taoist thought. In D. Munro (Ed.), *Individualism and holism: Studies in Confucian and Taoist values* (pp. 85–100). Ann Arbor: Center for Chinese Studies, University of Michigan.

Yang, G. (2003a). The Internet and civil society in China: A preliminary assessment. *Journal of Contemporary China, 12*(36), 453–475.

Yang, G. (2003b). The co-evolution of the Internet and civil society in China. *Asian Survey, 43*(3), 405–422.

Yang, S., & Song, J. (2004, October). *College students' self-positioning and the cell-phone consumption*. Paper presented at Mobile Communication and Social Change, Seoul, Korea.

Yuen, S. P., Law, P., & Ho, Y. Y. (2004). *Marriage, gender and sex in a contemporary Chinese village*. Armonk: M.E. Sharpe.

Zha, J. (1995). *China pop: How soap operas, tabloids, and bestsellers are transforming a culture*. New York: New Press.

Chapter 17
ICTD: Internet Adoption and Usage Amongst Rural Users in China

Jinqiu Zhao

1 Introduction

With both the largest population and the largest proportion of rural inhabitants in the world, China has witnessed unprecedented economic growth during the last decade, which has been reflected in its adoption of new technologies. After modest beginnings in the mid-1990s, Internet use has increased dramatically; the number of users[1] jumped from 0.62 million in 1997 to 298 million in December 2008. China is now ranked first in the world in terms of Internet users (CNNIC 1997, 2009).

However, the rapid diffusion of the technology since the late 1990s has been confined to the city centres, largely bypassing the majority of the population residing in the rural areas. The number of rural Internet users stood at 84.6 million by December 2008, accounting for approximately 28.4% of China's total Internet users (CNNIC 2009). Poor communication infrastructures, low levels of technological knowledge and limited incomes have put rural people at a disadvantage, resulting in a huge digital divide between the more developed coastal areas in the east and vast rural areas in the west.

Despite this discrepancy, the Internet is generally perceived as a new engine for rural empowerment, and a number of experimental projects have been initiated to test this view. This study explores the effects of Internet use on various aspects of rural development, including economic conditions and education.

[1] China Internet Network Information Centre (CNNIC) defines the Internet users as Chinese citizens aged 6 and above who have used the Internet over the last 6 months.

J. Zhao (✉)
Institute of Communication Studies, Communication University of China, Beijing, China
e-mail: zjq_99@yahoo.com

P.-l. Law (ed.), *New Connectivities in China: Virtual, Actual and Local Interactions*,
DOI 10.1007/978-94-007-3910-9_17,

2 ICTD Debate

The rapid development and dissemination of new communication technologies has generated a new wave of enthusiasm amongst developing countries; they see an opportunity to harness these advanced technologies in order to access the global information infrastructure and participate in the knowledge economy. This optimism has spread to encompass other, more basic, areas of potential development, such as ensuring a decent standard of living, improved health, better educational facilities and greater political participation.

The new arenas that have opened up as a result of the Internet have promised wider opportunities to alleviate poverty, enhance education, promote gender equality, improve health and encourage collective action for rural people in developing countries. However, the fact remains that, although ICTs have the potential to address the concerns of rural areas in developing countries, there is a widening gap between the developed and developing worlds and between the urban and rural areas in terms of the dissemination of ICTs and the distribution of their benefits.

As a consequence, the literature regarding ICTs and rural development is ambivalent. The optimists, including international organizations such as UNDP, World Bank and ITU, urge developing countries to make huge investments in their national information infrastructures so that their problems can be tackled with the aid of ICTs. The UNCSTD working group concluded that 'Although the costs of using ICTs to build national information infrastructures which can contribute to innovative "knowledge societies" are high, the costs of not doing so are likely to be much higher' (Mansell and Wehn 1998: 7).

However, these conclusions are countered by the pessimists, who argue that research findings connecting ICTs and economic growth only pertain to developed countries and that there is a lack of clear evidence of the same effect in developing countries. Anderson et al. (1999) note that technology is a means, not an end in itself, and the utilization of the Internet involves more than mere access. They warn that developing countries have not seemed to learn from disappointing experiences in technological transfer. More emphasis should be put on people, organizations and processes than on technologies per se. The pessimists maintain that money in developing countries would be better spent on elementary education and basic infrastructures such as transportation.

These conflicting views regarding ICTs and rural development are also cited in academic debates on the relationship between society and technology. Technological determinism (effectively the opposite of social determinism) views technology as a key governing force in society. Chandler (1995, Technological autonomy section, para. 1) observes, 'Rather than as a product of society and an integral part of it, technology is presented as an independent, self-controlling, self-determining, self-generating, self-propelling, self-perpetuating and self-expanding force'. The social determinism view, which has gained increasing recognition in recent years, underscores 'the "choices" (though not necessarily conscious choices) inherent in both the design of individual artefacts and systems, and in the direction or trajectory of innovation programmes' (Williams and Edge 1996: 857). The central issue here is whether the social context determines technological development or technology determines the social context, including its norms and values.

There is little empirical data on the relationship between the Internet and rural development (Quibria and Tschang 2001). One possible explanation is that the technology was introduced in many developing countries during the last 5 years, and it is still too early to gauge its social and economic impact. There are many inflated claims about the role of ICTs in poverty reduction and education promotion; research on the topic does not seem to differentiate between what the Internet can be expected to achieve and what has actually been achieved.

Consequently, there is a huge discrepancy between the claims of progress promised by ICTs and the empirical findings underpinning these claims. Furthermore, most of these empirical experiences are anecdotal and lack theoretical conceptualization and sound methodological design. Much more micro-level research is needed to understand and assess the impact of ICTs on the livelihood of the rural people. Such research would determine the extent of the access of the rural poor to Internet-based services and the degree to which these services actually generate incomes, promote education and health and enhance collective action.

3 Research Design

In order to achieve a comprehensive understanding of the phenomenon under investigation, the strategy of maximum variation sampling was used to select the research settings. The study was conducted in two rural areas—the town of Yellow Sheep River and Pinggu district. The Internet was introduced to Yellow Sheep River through the initiative of a private company and to Pinggu through the funding and support of local governments.

Data were collected from April to May 2005. The period spent in each rural site was around 20 days. Data collection methods included in-depth interviews, participant observations and documentation analysis. The strategies of maximum variation and snowball sampling were employed in the selection of the 29 interviewees. At each research site, government officials in charge of the planning and implementation of ICT projects were interviewed. The technicians in charge of Internet services were also approached for interviews. In turn, these informants recommended interviewees amongst local Internet users.

Participant observation focused on the daily experiences of farmers with Internet access and the activities at the information centres. Particular attention was paid to Internet access mode and use patterns, frequently used functions and websites and the influence of the Internet on the lives of rural people. In addition, the daily management of information centres and the provision of technical support offered were taken into consideration.

Public records, such as county-level censuses, statistics reports, county office records and government circulars, were used to gain a contextual understanding of the level of socioeconomic development and the rural telecommunication infrastructures. With regard to ICT projects carried out at Yellow Sheep River and Pinggu, internal records such as project implementation reports, minutes of meetings, presentations, logs of Internet access and ICT training materials were collected to supplement the data gathered through interviews and observations.

The interviews were transcribed, and field notes, field diaries and documents were typed up as word processing documents. The transcribed data of each research setting were treated as a separate data set so that comparisons could be made. The next stage was coding, which allowed the researcher to reduce the amount of data and focus on more meaningful analytical units. The 'constant comparative technique' was employed to identify and describe recurrent themes in the interviews (Wimmer and Dominick 2000: 107). After repeated reviews of the data, the material associated with certain topics were sorted out and labelled according to category. This required a constant refinement and elaboration of categories until an exact definition of each category and rules governing the inclusion of incidents were established. This process further narrowed the focus of the study.

Following the initial coding, a master list of all the codes developed and used in the study was created. The next step involved identifying relationships amongst categories and organizing them into different levels. At this stage, interrelationships and common regularities and patterns were derived across categories. Last, the themes were integrated into a conceptual schema that tied the observations about the transcribed data together to offer a logical interpretation of the phenomenon under study. The experiences and perceptions of the informants are reconstructed in the report of summarized findings.

4 Findings

4.1 The Town of Yellow Sheep River

Yellow Sheep River is an impoverished mountainous town in Gulang County, bordering the desolate Gobi Desert. About 23,000 rural people dwell in 81 settlements within 175 km^2, most of which are located in hard-to-reach mountain areas. Due to the formidable natural conditions and a dry climate, the town has a self-sufficient economy, dominated by agriculture, with traditional farming and livestock rearing as the chief means of making a living. The harvested crops go first to the daily food consumption. The remaining crops, if any, are sold in the local market where a barter trade is still practised. As 98.7% of the townspeople are engaged in agricultural production, the per capita annual income stood at less than 140 US dollars in 2004 (*Gulang Gaikuang* 2005).

The Internet came to this rural town as part of an experiment known as the 'Town and Talent Project' initiated by Wen Sayling, who was the late vice-chairman of a Taiwanese company. Wen set up the Town and Talent Technologies Cooperation in 2002 in the hope of using the Internet to eliminate poverty in underdeveloped areas.

As every rural town had at least one secondary school with teachers who were relatively well educated, schools were targeted as the means of promoting the use of new ICTs in these areas. In Yellow Sheep River, the project chose the vocational school to lead the experiment. Internet technology was primarily disseminated through two groups: the students and teachers on campus and scattered farming households off campus. As the first school involved in the project, the vocational middle school had received 140 donated computers from 2000 to 2004. Two computer classrooms were set up on campus to improve e-literacy amongst teachers and students.

Concomitant with the wide use of the Internet in teaching and learning, efforts were made to extend the benefits of the technology to the local farmers. To reinforce the school's leadership in fostering local economic development, the Town and Talent Project established an e-commerce centre on campus in the hope of encouraging technological awareness amongst farmers and leading to more profitable methods of farming and selling. The e-commerce centre, equipped with 25 computers, a printer, a digital camera and a fax machine, was registered as a business under the auspices of the Town and Talent Project. Town and Talent Technologies supervised the daily operation of the centre and offered technical support whenever necessary.

The local farmers could learn how to use a computer at the e-commerce centre. Information on farming skills, land policies and prices of seeds and fertilizers was posted on the walls when the weather was fine. People would usually stop to read the postings when they came for the regular town fairs. The off seasons for farming were the busiest times for the centre. Since opening in 2000, it organized two dozen Internet training courses for farmers, as well as 36 training sessions tailored to the special needs of the village chiefs, entrepreneurial farmers and students from the schools nearby. The most frequently taught courses were those providing basic knowledge about computer hardware, popular software and Internet surfing.

The centre was frequented by both teachers and local farmers. It did not charge teachers for their Internet use as the Internet service was considered part of the teaching facilities. Farmers were charged 2 *yuan* per hour to use the service. The running costs of the centre (including Internet access, electricity, phone and hardware replacement) were covered by the tuition for the training sessions, fees from daily users and the 1% commission charged for every transaction completed via the e-commerce service.

The chief administrator of the centre was Hu Wanlong, a history teacher at the vocational school. As a part-time administrator and manager of the e-centre, Hu helped farmers by conducting searches, offering technical support, maintaining the hardware and keeping the centre financially viable. In addition to managing the centre, Hu also served as an informal advisor to the farmers on issues ranging from the new types of tractors acquired in neighbouring villages to the latest government policies on land and farming. Despite the training provided at the e-centre, few local farmers could actually operate the computers and access the Internet by themselves. Their use of the Internet was usually mediated by Hu, who became a central figure because most social changes that the town experienced revolved around the e-commerce centre during the period of his management.

The e-commerce centre was mainly devoted to improving the traditional methods of cultivation and selling. The farmers were more willing to accept the concept of scientific farming after visiting various agricultural websites at the training sessions. These websites offered useful information on scientific advancements in land fertilization and weed and insect control. Furthermore, many financial transactions, such as the selling of green peas, yams and celery, were accomplished by posting information about the local farm produce online.

Although there were positive changes brought about by the Internet, the technology was not well received by the local farmers. Those who benefited most from the Internet services were either Hu's acquaintances or villagers who became familiar with the e-centre. Even 5 years after the introduction of the Internet, there were no fundamental changes in the quality of life of the rural people in this remote town. The annual per capita income of local villagers increased slightly, from 1,060 *yuan* (US $133) in 2000 to 1,123 *yuan* (US $140) in 2004. Hu reported that the total income of the e-centre was about 3,000 *yuan* (US $375) in 2004, which was just enough to cover the costs of Internet access (200 *yuan* or US $25 per month), electricity, phone and hardware replacement. Hu did not expect the e-centre to become profitable because neither the number of Internet users nor the frequency of online transactions was likely to rise significantly.

4.2 Pinggu District

Pinggu is a suburban district in the northeast of the municipality of Beijing. With an area of 1,075 km^2 and a population of 400,000, Pinggu contains 16 townships and 275 villages. The district began to expand its ICT infrastructure towards the end of 1998, when the local government set up an information office to provide Internet support for government institutions and local enterprises. In June 2000, the local government initiated the agricultural information network project to provide Internet connections for farmers. Tao*yuan* village in the town of Zhenluoying and Nandingfu village in the town of Machangying were the two pilot villages selected to carry out the experiment.

4.3 Taoyuan Village

Taoyuan is a remote mountainous village in the far north of Pinggu. In Taoyuan, 97 households, comprising 327 residents, live in an area of 3.38 km^2. The village specializes in fruit shipping and marketing: over 80% of the households are engaged in storing and selling fruits, and the average earnings of fruit dealers have increased to 6,000 *yuan* (US $750) per year.

In order to facilitate farmers' access to market information and help them enlarge the scale of their operations, the information office established by the Pinggu government provided Internet access to Taoyuan in 2001. In view of the scattering of rural households in the mountains, the information office did not set up a central location to offer collective Internet access. Instead, it encouraged individual households to buy computers and log onto the Internet from home. With partial financial support from the district government and the village committee, over 30 families bought self-assembled computers at a price of 4,000 *yuan* (US $500), half the cost of name-brand products.

Because the local telecommunication operators did not provide the Internet connection service, the information office, together with a cable TV company, made use of the existing cable TV network to set up Internet access for the rural households. A cable modem was installed at the subscriber end, which allowed a computer to be connected with the Internet. The cable access enabled farmers to enjoy high-speed Internet service.

To help the farmers develop computer and Internet application skills, the information office dispatched technical staff to offer basic e-literacy training. About 40 farmers acquired skills in typing, computer operation and searching online. In 2001, the information office helped Taoyuan set up a village website (http://taoyuan.bjpg.gov.cn) to promote their fruit and facilitate business transactions. The Internet exposed Taoyuan farmers to innovative ways of selling. Online communication saved them the trouble of travelling great distances to do business; many clients contacted local fruit dealers by calling the phone numbers listed on the website. The farmers also used the Internet to keep track of fruit prices at the region's wholesale market so they could ensure that their prices were competitive.

During the 2 years following the introduction of the Internet to Taoyuan, the technology was used intensively by local farmers. The village committee covered the Internet subscription costs by paying a monthly flat rate of 50 *yuan* (US $6.25) for each household that was part of the network. Business contacts were established with more than 20 enterprises across the country. Over 1,400 tons of fruits were sold, and the farmers in the village earned 800,000 *yuan* (US $100,000) in total.

However, the farmers' enthusiasm for the Internet began to slacken in 2004. Two-thirds of the households that acquired Internet access sold their computers. This is partly attributable to the fact that the village stopped paying the monthly subscription fees for the farmers in early 2004. The head of the village explained that costs for Internet connection had become a huge financial burden. With no village enterprises and collective revenue, the village did not have the money to pay for the access fees, which totalled 18,000 *yuan* (US $2,250) per year.

Moreover, the Internet service provided by the cable company was not well maintained. When the cable was disabled due to thunderstorms, it sometimes took months before the cable company attended to the problem. Owing to the unprofitable Internet subscription rate in the mountainous region, the cable company was not motivated to provide Internet services. The village was also receiving less technical support from the information office, which had become overextended as more and more government institutions went online in Pinggu.

4.4 Nandingfu Village

Nandingfu village is under the direct administration of the town of Machangying, situated in the west of Pinggu. The village has 235 rural households, comprising 610 residents, within about 73 ha of arable land. As a result of robust development of the village economy in recent years, the collective revenue of Nandingfu reached over four million *yuan* (US $500,000) and local farmers' per capita annual income stood at 8,100 *yuan* (US $1,012) in 2004.

Benefiting from the strong economy, Nandingfu is far ahead of other rural communities in terms of the development of basic infrastructures and the provision of social welfare. In order to provide farmers with opportunities to use modern communication technologies, Nandingfu decided to promote e-literacy amongst local farmers at the end of 2002. The initiative received the full support of the information office of the Pinggu government, which undertook to find a viable model of disseminating the Internet amongst the rural residents. Given that Nandingfu could afford the potential costs associated with the Internet, the information office suggested that an information centre should be established to offer public Internet access.

With partial financial support from the district government, Nandingfu invested 50,000 *yuan* (US $6,250) in 19 computers and 50 desks and chairs. Fifteen computers were for use in the information centre, and the rest were employed by the village committee for accounting and daily management purposes. The information centre is located next to the office of the village committee, which is within walking distance for all the households.

In the early days of the information centre, the village often invited technicians from the information office to deliver lectures. Anyone with an interest was given an opportunity to learn. After a couple of training sessions, around 60 young farmers had acquired Internet skills. A female farmer became a part-time manager of the information centre.

Unlike other e-centres, the information centre at Nandingfu does not provide information and communication services to the villagers. Instead, it is used by the town government to provide training in technical literacy. Because it is the only one of its kind in the area, the centre has been extensively used to promote technical literacy amongst farmers in surrounding villages under the direct administration of Machangying. The chief organizers of training courses often represent various branches of the town government, such as the agricultural or judicial departments, women's association and youth organization. Farmers enrol in these free courses on a voluntary basis. A typical training course lasts 3 months and includes both formal instruction and hands-on exercises.

The manager of the information centre, Bao, said that most of the learners were 25–40 years old and about 80% were female. Women were strongly represented because they tended to be less busy than men and, in many cases, wanted to learn these skills so they could educate their children. With the e-literacy skills, they have learned at the training courses, some farmers have set up their own businesses, offering photocopying and typing services, and a few have bought their own computers so they can surf the Internet at home. According to Bao, while these new technologies are unlikely to play a major role in rural people's life at present, the e-literacy training helps to create awareness of the possibilities afforded by the technology and to foster the skills necessary for Internet use. Without these skills, farmers would be easily bypassed by technological advances.

The centre was connected to the Internet by dial-up until 2004 when the district government invested 10,000 *yuan* (US $1,250) to extend broadband service to the village. Since the establishment of the information centre, all the costs of hardware maintenance and Internet access have been covered by the village committee. The village head, Li Guang'en, said that the Internet subscription fees were around 4,000 *yuan* (US $500) per year.

Following its experiment at the two pilot villages, Pinggu set up 12 information centres within 4 years that serviced half the towns in the district. The information centres have trained 5,600 information technicians at the district, township and village levels and 38,000 farmers. By the end of 2005, 5,773 fibre kilometres of optic cable had been installed to link up every village in the district (CRTDC 2005).

5 The Effects of the Internet on Rural Development

Despite inflated claims regarding its impact and the high expectations it raises, the Internet does not always bring about drastic social and economic advances, as the cases covered in this study demonstrate. Overall, there are positive signs of social and economic progress as a result of the diffusion and use of the Internet in the two areas under study. However, the scope and the depth of the changes brought about by the technology are not significant enough to lend support to the optimistic visions presented in the literature regarding the role of ICTs in general and the Internet in particular in promoting development. Internet diffusion and use have not resulted in fundamental transformations of modes of production, organization of work and norms of consumption in rural society.

Though the Internet did not bring the expected changes to these rural areas, the empirical findings of this study support the positive effects of the Internet on rural development. The following discussion focuses on the effects that the Internet has had on the human development of rural users. As economic concerns have been the primary motivation for introducing the Internet to rural communities, the effects of its introduction have been mainly felt in the economic sector of rural life.

The findings show that the Internet contributes to agricultural productivity by increasing farmers' access to advanced agricultural technologies and facilitating an exchange of information on farming innovations. The market information and analysis provided on agricultural websites assists farmers in making rational decisions and aligning farm output to market supply and demand. For example, the price information on the web enabled a few farmers in a neighbouring county of Yellow Sheep River to set a reasonable market price for the medicinal herbs they collected from mountains, whereas, previously, they had to accept the prices offered by middlemen.

Rural residents view the Internet as an efficient channel of communication to promote their farm produces and establish business contacts. In every rural centre covered in this study, websites were set up to market local goods, attract investment and enlarge business scales. These websites not only serve as windows through which rural people can look beyond their immediate milieu to seek business opportunities, they also allow rural markets to be known to the rest of the world and incorporated into the digital network.

The village website in Taoyuan provided a business platform for local farmers to advertise their fruit and carry out transactions over the Internet. Online sales saved Taoyuan farmers the effort of travelling great distances to seek potential trade partners, and thereby greatly reduced the costs of sales. Within 2 years of adoption of the Internet, local fruit dealers managed to extend their businesses to many parts of the country and establish long-term relationships with over 20 enterprises. The increase in orders propelled rural residents to unite in order to establish organized sales of marketable surplus and bulk transport of fruit. E-commerce created more business opportunities and led to a significant rise in average household earnings.

In terms of its effects on rural education, the Internet has been used to enhance teaching quality and learning capabilities at rural schools and to increase the level of technological readiness amongst adults. Yellow Sheep River achieved impressive results as a consequence of using computers and the Internet to improve middle-school education; Pinggu's success in training farmers in e-literacy was equally inspiring.

The vocational middle school in Yellow Sheep River witnessed the marked changes brought about by the introduction of the Internet to the educational programme. Once they had a grasp of computer operational skills, the teachers were able to access the vast amount of educational resources available on the Internet. Through visiting the websites of elite middle schools in large cities, the teachers could improve their teaching methods without the necessity of actually visiting the schools themselves. Moreover, active Internet use was conducive to creating an academic atmosphere and allowed schools to keep up with current teaching and learning methods.

With an average of five to six students sharing one PC, the school enjoyed the best computer facilities in town. As young people are amongst the quickest to grasp new technologies, e-literacy acquired at an earlier age has a long-lasting impact. Even though the students were only allowed to browse the school network, they were able to read updated news and information about the outside world, which helped to enlarge their horizon and gradually shape their perceptions. The Internet stirred up a strong desire to improve their lot, and the students anticipated a better life by acquiring knowledge and information. With the widespread use of computer and network technologies on campus, the school showed a steady increase in enrolment and in the number of graduates who went on to pursue higher education.

The change agencies involved in introducing the Internet to rural areas made persistent efforts to provide rural people with opportunities of learning computer and Internet application skills. Over 60 training sessions were organized at the e-centre in Yellow Sheep River to help entrepreneurial farmers, village chiefs and students grasp basic computer knowledge. The most systematic and large-scale e-literacy programme was carried out at the information centre in Nandingfu village in Pinggu. During a period of 4 years, the village collaborated with 11 other information centres in the district, to provide over 44,000 farmers and information technicians, representing 14% of the population of the area, with formal training.

The real changes that can be accomplished as a result of one training session remained doubtful; only a small percentage of the farmers participating in the programme were able to operate computers and surf the Internet after the training. Still, the promotion of e-literacy, together with the extension of the Internet technologies, has had a profound effect on narrowing the digital divide between urban and rural areas. The extension of the Internet technologies has increased the chances for rural people to acquire the knowledge that this technology provides; without it, the majority would be totally deprived of opportunities to know the benefits associated with advanced ICTs.

E-literacy training has also given rise to a number of information technicians, innovators and early adopters in rural areas. Given the severe shortage of technical personnel in rural areas, these residents have the opportunity to become the backbone of rural communities in terms of providing technical support and maintenance services to governments, private organizations and individuals. Generally speaking, the introduction of the Internet into rural communities has accomplished some important objectives: it has enhanced capacity building, increased the level of e-readiness and fostered an environment conducive to the future employment of ICTs.

6 Conclusion

The experiences of the rural areas discussed in this study indicate that Internet diffusion and use must be compatible with the existing political, socioeconomic and technological contexts of the rural settings if it is to serve as an engine for development. The interplay of structural factors and individual characteristics of each area very much affects the diffusion and use of the Internet and, consequently, the social and economic benefits accrued.

Amongst the structural factors identified by this study, the organizational forces play a dominant role during the entire diffusion process. The Internet does not come to the rural people as a natural stage in the development of the economy and infrastructure of rural areas; instead, its diffusion is undertaken by the change agencies, who hope that the new technology will open up opportunities. In the context of rural China, where local economies and infrastructures can hardly sustain such an advanced technology as the Internet, the adoption of the Internet is usually the result of organizational intervention. The change agencies that establish these programmes play an essential role in extending the technology to the rural people. The reasons motivating the introduction of these programmes play a large role in determining the means of Internet access provided, the efforts made to encourage farmers to adopt and use the technology, and the direction of social changes.

However, despite the high hopes the change agencies have pinned upon the Internet to bring about intended changes, the degree to which the Internet can be adopted and used by rural people is conditioned by the joint influence of structural factors and individual characteristics. Individual Internet use is greatly determined by the user's technical abilities, information needs and appreciation of perceived benefits; both needs and benefits are, in turn, determined by production modes, the relevance of web content and the existing information channels.

One of the unexpected findings of this study is that Internet adoption and use tend to be closely associated with local farming and business activities. Information needs, conditioned by modes of production, motivate farmers to seek information actively. Farmers are therefore inclined to adopt Internet technology. In other words, impoverished and conservative areas where self-sufficiency is the dominant economic mode are less likely to rely on the Internet than rural areas with a relatively developed economy and a strong market orientation.

After the adoption of the Internet, its continued use depends on the availability of relevant web content that meets farmers' daily information needs and its level of convenience in comparison with other modes of receiving market information. Early research findings on the diffusion of new media technologies stress the part played by the information needs of potential adopters (Bowden et al. 1994; Mullaly-Quijas et al. 1994; Wallingford et al. 1996). The findings of this study show that the degree to which web content satisfies information needs influences potential adopters' evaluation of the Internet, as well as their use of the technology in the post-adoption stage.

However, the effectiveness of the Internet in meeting the information needs is not solely determined by the relevance of web content. Prior to the introduction of new ICTs, rural communities relied on an existing network of old ICTs or interpersonal communication. Therefore, the availability and relevance of the web content, the cost-effectiveness of Internet adoption and usage and the relative advantage of the Internet as compared to the existing media channels are amongst the key factors that shape Internet adoption and usage. These factors all play a part in motivating rural individuals to experiment with the Internet, to adapt their behaviours to the new technology or to discontinue use.

This study also highlights the role of the key stakeholders involved in Internet diffusion and use in the rural areas. The deployment of the Internet to promote rural development depends very much on the perceptions of its benefits on the part of rural administrators, telecommunication operators, private investors and rural individuals.

In summary, this study shows how social structures exert influence on Internet diffusion, adoption and use. The findings of the study have provided strong empirical evidence in support of the argument of the social shaping of technology. The Internet serves as an agent of change in rural areas, but the extent, duration and intensity of its impact are primarily determined by existing socioeconomic conditions.

References

Anderson, J., Crowder, L. V., & Dion, D. (1999). Applying the lessons of participatory communication and training to rural telecentres. From http://www.fao.org/WAICENT/FAOINFO/SUSTDEV/CDdirect/CDan0010.htm. Accessed 15 Nov 2004.

Bowden, V. M., Kromer, M. E., & Tobia, R. C. (1994). Assessment of physicians' information needs in five Texas counties. *Bulletin of the Medical Library Association, 82*(2), 189–196.

Chandler, D. (1995). Technological or media determinism. From http://www.aber.ac.uk/media/Documents/tecdet/tdet06.html. Accessed 9 Feb 2006.

China Internet Network Information Centre. (1997, October). Statistical report on the development of Chinese Internet. From http://www.cnnic.net.cn/develst/9710/e-9710.shtml. Accessed 14 Apr 2003.

China Internet Network Information Centre. (2009, January). The 23rd statistical survey report on the Internet development in China. From http://www.cnnic.net.cn/uploadfiles/pdf/2009/1/13/92458.pdf. Accessed 16 Mar 2009.

China Rural Technology Development Centre. (2005). *Jiaqiang nongcun xinxihua jianshe cujin pinggu jingji fazhan* [Promoting the economy of Pinggu through building rural information network]. From http://www.crst.cn/infor/pinggu.htm. Accessed 1 Jan 2006.

Gulang gaikuang. (2005). *Gulang gaikuang* [The overview of Gulang County]. From http://www.gulang.gansu.gov.cn/zoujin/view.asp?NewsID=247&classID=14. Accessed 30 June 2005.

Mansell, R., & Wehn, U. (Eds.). (1998). *Knowledge societies: Information technology for sustainable development*. New York: Oxford University Press Inc.

Mullaly-Quijas, P., Ward, D. H., & Woelfl, N. (1994). Using focus groups to discover health professionals' information needs: A regional marketing study. *Bulletin of the Medical Library Association, 82*(3), 305–311.

Quibria, M. G., & Tschang, T. (2001, January). Information and communication technology and poverty: An Asian perspective. From http://www.globalknowledge.org/gkps_portal/aprm2002/workshop/ADBI.pdf. Accessed 22 Nov 2004.

Wallingford, K. T., Ruffin, A. B., Ginter, K. A., Spann, M. L., Johnson, F. E., Dutcher, G. A., et al. (1996). Outreach activities of the National Library of Medicine: A five-year review. *Bulletin of the Medical Library Association, 84*(2), 1–60.

Williams, R., & Edge, D. (1996). What is the social shaping of technology? *Research Policy, 25*, 856–899.

Wimmer, R. D., & Dominick, J. R. (2000). *Mass media research: An introduction* (6th ed.). Belmont: Wadsworth Publishing Company.

Index

P.-l. Law (ed.), *New Connectivities in China: Virtual, Actual and Local Interactions*,
DOI 10.1007/978-94-007-3910-9, © Springer Science+Business Media B.V. 2012

MIX
Papier aus verantwortungsvollen Quellen
Paper from responsible sources
FSC® C105338

If you have any concerns about our products,
you can contact us on
ProductSafety@springernature.com

In case Publisher is established outside the EU,
the EU authorized representative is:
Springer Nature Customer Service Center GmbH
Europaplatz 3, 69115 Heidelberg, Germany

Printed by Libri Plureos GmbH
in Hamburg, Germany